Pressure Vessels Field Manual

Pressure Vessels Field Manual

Common Operating Problems and Practical Solutions

Maurice Stewart
Oran T. Lewis

ELSEVIER

AMSTERDAM • BOSTON • HEIDELBERG • LONDON
NEW YORK • OXFORD • PARIS • SAN DIEGO
SAN FRANCISCO • SINGAPORE • SYDNEY • TOKYO

Gulf Professional Publishing is an Imprint of Elsevier

G P
P

Gulf Professional Publishing is an imprint of Elsevier
225 Wyman Street, Waltham, MA 02451, USA
The Boulevard, Langford Lane, Kidlington, Oxford, OX5 1GB, UK

Library of Congress Cataloging-in-Publication Data
Stewart, Maurice.
Pressure vessels field manual: common operating problems and practical solutions / Maurice Stewart and Oran T. Lewis.
p. cm.
Includes bibliographical references.
ISBN 978-0-12-397015-2
1. Pressure vessels–Maintenance and repair–Handbooks, manuals, etc. 2. Pressure vessels–Maintenance and repair–Standards–United States–Handbooks, manuals, etc. 3. Pressure vessels–Inspection–Handbooks, manuals, etc. 4. Pressure vessels–Inspection–United States–Handbooks, manuals, etc. 5. Boilers–Maintenance and repair–Handbooks, manuals, etc. 6. Boilers–Maintenance and repair–Standards–United States–Handbooks, manuals, etc. 7. Boilers–Inspection–Handbooks, manuals, etc. 8. Boilers–Inspection–United States–Handbooks, manuals, etc. 9. ASME boiler & pressure vessel code. I. Lewis, Oran T. II. Title.
TS283.S74 2012
681'.76041–dc23
2012026721

British Library Cataloguing-in-Publication Data
A catalogue record for this book is available from the British Library.

For information on all Gulf Professional Publishing publications visit our website at http://store.elsevier.com

09 10 11 12 10 9 8 7 6 5 4 3 2 1

Printed in the United States of America

Contents

History and Organization of Codes

1

▶ OVERVIEW

This chapter discusses the following topics:

Use of pressure vessels and equipment
History of pressure vessel codes in the United States
Organization of the American Society of Mechanical Engineers (ASME) Boiler and Pressure Vessel Code
Updating and interpreting the code
ASME Code stamps
Organization of the American National Standards Association (ANSI)/ASME B31 Code for Pressure Piping
Some other pressure vessel codes and standards in the United States
Worldwide pressure vessel codes
ASME Code, Section VIII, Division 1 versus Division 2 and Division 3
Design criteria, ASME Code, Section VIII, Division 1
Design Criteria, ASME Code, Section VIII, Division 2
Welding criteria, ASME Code, Section IX
ASME Code, Section I, Power Boilers
Additional requirements employed by users in critical services

▶ USE OF PRESSURE VESSELS AND EQUIPMENT

Process vessels are used at all stages of processing oil and gas
 Initial separation
 Processing, conditioning, and treating
 Storage
Pressure vessels are made in all sizes and shapes

- Less than an inch and greater than 150 feet in diameter
- Buried in the ground or deep in the ocean
- Most are positioned on the ground or supported on platforms

Internal pressures
- Less than 1 inch water gauge pressure or greater than 300,000 psi
- Normal range for monoblock construction is between 15 to 5000 psi

ASME Boiler and Pressure Vessel Code
- Specifies a minimum internal pressure from 15 psi
- May require special design requirements at internal pressures above 3000 psi
- Any pressure vessel that meets all the requirements of the ASME Code, regardless of the internal or external design pressure, may still be accepted by the authorized inspector and stamped by the manufacturer with the ASME Code symbol
- American Petroleum Institute (API) storage tanks
- May be designed for and contain no more internal pressure than that generated by the static head of fluid contained in the tank

HISTORY OF PRESSURE VESSEL CODES IN THE UNITED STATES

In the late 1800s and early 1900s, explosions in boilers and pressure vessels were frequent
- Firetube boiler explosion on the Mississippi River steamboat *Sultana* on April 27, 1865, resulted in the boat sinking within 20 minutes and killing 1500 people
- Firetube boiler explosion in a shoe factory in Brockton, Massachusetts, in 1905 killed 58, injured 117, and did $400,000 in property damage

- Another explosion in a shoe factory in Lynn, Massachusetts, in 1906 resulted in death, injury, and extensive property damage
- Massachusetts governor directed the formation of a Board of Boiler Rules
- First set of rules approved in 1907
- The code was three pages long!

In 1911, the ASME Boiler and Pressure Vessel Code Committee was established to perform the following functions:

- Formulate minimum rules for the construction of boilers
- Interpret the rules when requested
- Develop revisions and additional rules as needed

In 1915, the first ASME Boiler Code was issued:

- Titled "Boiler Construction Code, 1914 Edition"
- Beginning of the various sections of the code, which ultimately became Section I, "Power Boilers"

In 1925, the first ASME Code of Pressure Vessels was issued:

- Titled "Rules for the Construction of Unfired Pressure Vessels," Section VIII
- Applied to vessels over 6 inches in diameter, volume over 1.5 ft^3, and pressure over 30 psi

In 1931, a joint API-ASME committee was formed to develop an unfired pressure vessel code for the petroleum industry

In 1934, the first pressure vessel code for the petroleum industry was issued, and for the next 17 years, two separate unfired pressure vessel codes existed

In 1952, the two codes were consolidated into one, titled "ASME Unfired Pressure Vessel Code," Section VIII

In 1968, the original code became Section VIII, Division 1, "Pressure Vessels," and another part was issued, which was Section VIII, Division 2, "Alternate Rules for Pressure Vessels"

ANSI/ASME Boiler and Pressure Vessel Code is issued by the ASME with the approval by the ANSI as an ANSI/ASME document

ANSI/ASME Boiler and Pressure Vessel Code has been established as the set of legal requirements in many countries of the world

Most piping systems are built to the ANSI/ASME Code for Pressure Piping B31; there are a number of different piping code sections for different types of systems
- Section I uses B31.1
- Section VIII uses B31.3

▸ ORGANIZATION OF THE ASME BOILER AND PRESSURE VESSEL CODE

ASME Boiler and Pressure Vessel Code is divided into

Sections
Divisions
Parts
Subparts

Some of these sections relate to
- A specific kind of equipment and application
- Specific materials and methods for application and control of equipment
- Care and inspection of installed equipment

At the present time, the code includes the following sections.

Section I: Power Boilers

This section provides requirements for all methods of construction of power, electric, and miniature boilers;

high-temperature water boilers used in stationary service; and power boilers used in locomotive, portable, and traction service.

Rules pertaining to the use of the V, A, M, PP, S, and E Code symbol stamps are also included. The rules are applicable to boilers in which steam or other vapor is generated at a pressure exceeding 15 psig and high-temperature water boilers intended for operation at pressures exceeding 160 psig or temperatures exceeding 250°F.

Superheaters, economizers, and other pressure parts connected directly to the boiler without intervening valves are considered as part of the scope of Section 1.

This section is briefly reviewed in this text.

Section II: Materials

Part A: Ferrous Materials Specifications

This part is a service book to the other code sections, providing material specifications for ferrous materials adequate for safety in the field of pressure equipment. These specifications contain requirements and mechanical properties, test specimens, and methods of testing. They are designated by SA numbers and are derived from American Society of Testing Materials (ASTM) "A" specifications.

Part B: Nonferrous Material Specifications

This part is a service book to the other code sections, providing material specifications for nonferrous materials adequate for safety in the field of pressure equipment. These specifications contain requirements for heat treatment, manufacture, chemical composition, heat and product analyses, mechanical test requirements and mechanical properties, test specimens, and methods of testing. They are designated by SB numbers and are derived from ASTM "B" specifications.

Part C: Specifications for Welding Rods, Electrodes, and Filler Metals

This is a service book to the other code sections, providing material specifications for the manufacture, acceptability, chemical composition, mechanical usability, surfacing, testing requirements and procedures, operating characteristics, and intended uses for welding rods, electrodes, and filler metals. These specifications are designed by SFA numbers and are derived from American Welding Society (ANS) specifications.

Section III: Rules for Construction of Nuclear Power Plant Components

This section provides requirements for the materials, design, fabrication, examination, testing, inspection, installation, certification, stamping, and overpressure protection of nuclear power plant components and component and piping supports. Components include metal vessels and systems, pumps, valves, and core support structures. The components and supports covered by this section are intended to be installed in a nuclear power system that produces and controls the output of thermal energy from nuclear fuel and those associated systems essential to the functions and overall safety of the nuclear power system.

This section also provides requirements for (1) containment systems and transport packaging for spent fuel and high-level radioactive waste and (2) concrete reactor vessels and containments.

Section IV: Heating Boilers

This section provides requirements for design, fabrication, installation, and inspection of steam generating boilers and hot water boilers intended for low-pressure service that are directly fired by oil, gas, electricity, or coal.

It contains appendices that cover approval of new material, methods of checking safety valve and safety relief valve capacity, examples of methods of checking safety valve and safety relief valve capacity, examples of methods of calculation and computation, definitions relating to boiler design and welding and quality control systems.

Rules pertaining to use of the H, HV, and HLW Code symbol stamps are also included.

Section V: Nondestructive Examination

This section contains requirements and methods for nondestructive examination that are required by other code sections. It also includes manufacturer's examination responsibilities, duties of authorized inspectors, and requirements for qualification of personnel, inspection, and examination.

Examination methods are intended to detect surface and internal discontinuities in materials, welds, and fabrication parts and components. A glossary of related terms is included.

Section VI: Recommended Rules for the Care and Operation of Heating Boilers

This section covers general descriptions, terminology, and operation applicable to steel and cast iron boilers limited to the operating ranges of Section IV: Heating Boilers. It includes guidelines for associated controls and automatic fuel-burning equipment.

Section VII: Recommended Guidelines for the Care of Power Boilers

The purpose of these guidelines is to promote safety in the use of stationary, portable, and traction type heating boilers. This section provides such guidelines to assist operators of power boilers in maintaining their plants. Emphasis has been placed on individual-type boilers because of their extensive use.

Section VIII

Division 1: Pressure Vessels

This division of Section VIII provides requirements applicable to the design, fabrication, inspection, testing, and certification of pressure vessels operating at either internal or external pressures exceeding 15 psig.

Such pressure vessels may be fired or unfired. Specific requirements apply to several classes of material used in pressure vessel construction and also to fabrication methods such as welding, forging, and brazing.

It contains mandatory and nonmandatory appendices detailing supplementary design criteria, nondestructive examination, and inspection acceptance standards. Rules pertaining to the use of the U, UM, and UV Code symbol stamps are also included.

Division 2: Alternate Rules

This division of Section VIII provides requirements applicable to the design, fabrication, inspection, testing, and certification of pressure vessels operating at either internal or external pressures exceeding 15 psig. Such vessels may be fired or unfired. This pressure may be obtained from an external source or by the application of heat from a direct source, an indirect source, or any combination thereof.

These rules provide an alternate to the minimum requirements for pressure vessels under Division 1 rules. In comparison to the Division 1, Division 2 requirements on materials, design, and nondestructive examination are more rigorous; however, higher design stress intensity values are permitted.

Division 2 rules cover only vessels to be installed in a fixed location for a specific service where operation and maintenance control is retained during the useful life of the vessel by the user who prepares or causes to be prepared the design specifications.

These rules may also apply to human occupancy pressure vessels, typically in the diving industry. Rules pertaining to the use of U2 and UV Code symbol stamps are also included.

Division 3: High-Pressure Vessels

This division of Section VIII provides requirements applicable to the design, fabrication, inspection, testing, and certification of pressure vessels operating at either internal or external pressures generally above 10,000 psi. Such vessels may be fired or unfired. This pressure may be obtained from an external source, a process reaction, by the application of heat from a direct or indirect source, or any combination thereof.

Division 3 rules cover vessels intended for a specific service and installed in a fixed location or relocated from worksite to worksite between pressurizations. The user who prepares the design specifications or causes them to be prepared retains the operation and maintenance control during the useful life of the vessel.

Division 3 does not establish maximum pressure limits for Section VIII, Divisions 1 or 2, nor minimum pressure limits for this division. Rules pertaining to the use of the UV3 Code symbol stamps are also included.

Divisions 1 and 2 will be reviewed in detail in this text.

Section IX: Welding and Brazing Qualifications

This section contains rules relating to the qualification of welding and brazing procedures as required by other code sections for components manufacture.

It also covers rules relating to the requalification of welders, brazers, and welding and brazing operators in order that they may perform welding or brazing as required by other code sections in the manufacture of components.

Welding and brazing data cover essential and nonessential variables specific to the welding or brazing process used.

Section X: Fiber-Reinforced Plastic Pressure Vessels

This section provides requirements for construction of a fiber-reinforced plastic (FRP) pressure vessel in conformance with a manufacturer's design report. It includes the production, processing, fabrication, inspection, and testing methods required for the vessel.

Section X includes two classes of vessel design: Class 1, a qualification through the destructive test of a prototype, and Class II, mandatory design rules and acceptance testing by nondestructive methods.

These vessels are not permitted to store, handle, or process lethal fluids. Vessel fabrication is limited to the following processes: bag molding, centrifugal casting and filament winding, and contact molding. General specifications for the glass and resin materials and minimum physical properties for the composite materials are given.

Section XI: Rules for In-Service Inspection of Nuclear Power Plant Components

This section contains Division 1 and 3 in one volume and provides rules for the examination, in-service testing and inspection, and repair and replacement of components and systems in light-water-cooled and liquid-metal-cooled nuclear power plants.

The Division 2 rules for inspection and testing of components of gas-cooled power plants were deleted in the 1995 edition. With the decommissioning of the only gas-cooled reactor to which these rules apply, there is not apparent need to continue publication of Division 2.

Application of this section of the code begins when the requirements of the construction code have been satisfied. The rules of this section constitute requirements to maintain the nuclear power plant while in operation and to return the plant to service following plant outages and repair or replacement activities.

Section XII: Rules for the Construction and Continued Service of Transport Tanks

This section covers requirements for construction and continued service of pressure vessels for the transportation of dangerous goods via highway, rail, air, or water at pressures from full vacuum to 3000 psig and volumes greater than 120 gallons. This section also contains modal appendices containing requirements for vessels used in specific transport modes and service applications. Rules pertaining to the use of the T Code symbol stamp are also included.

Organization of Code Committee

Began with 7 members, currently there are more than 450 members
Composed of main committee, executive committee, conference committee and staff
"Book" committees: Sections I, III, IV, VIII, X
Special subcommittees
Section II: Materials
Section V: Nondestructive Examination (NDE)
Section IV: Welding
Section X : Nuclear In-Service Inspection
Service committees
Design
Safety valve requirements
Boiler and pressure vessel accreditation
Nuclear accreditation

▸ UPDATING AND INTERPRETING THE CODE

Addenda

Issued once a year on January 1
Permission at the date of issuance and become mandatory 6 months after that date

New Edition of the Code

Issued every 3 years on July 1 and becomes mandatory at that time
Incorporates all addenda to the previous edition
Does not incorporate anything new beyond that contained in the previous addenda except for editorial corrections or a change in the numbering system

Code Cases

Contain rules for materials and special constructions that have not been sufficiently developed for inclusion in the code itself

Issued four times a year

Annulled after a period of time unless renewed

Code Interpretations

Responds to inquires about the intent of the code

After a decision has been reached, it is forwarded to the inquiring party and also published in *Mechanical Engineering* magazine

If no further criticism is received, the decision may be formally adopted as a code interpretation

Issued every 6 months (refer to the appendix for an example)

In the form of questions and replies that further explain items in the code that have been misunderstood

May be included in the next addendum

Fifty percent of the interpretations come from Europe and Southeast Asia

Turnaround time is normally 1 month and 6 months on inquiries requiring research

Inquiries should be sent to:

Secretary of ASME Boiler Committee
345 East 47th Street
New York, NY 10017 USA

Or

ASME.org

▶ ASME CODE STAMPS

Most oil and gas company policies dictate that pressure vessels conform to the ASME Code requirements whether the authority having jurisdiction requires its use or not

Figure 1.1 UG-116 official symbols for stamp to denote the ASME standard.

"U" (Unfired Pressure Vessels) and "UM" (Unfired Miniature Pressure Vessels) (refer to Figure 1.1)

Section VIII, Division 2, provides the "U2" stamp for vessels built according to its requirements

Other sections use different stamps according to the type of equipment under their jurisdiction

"E" Electric Boilers (Section I)

"S" Power Boilers (Section I)

"N" Nuclear Vessels (Section III)

"H" Heating Boilers (Section IV)

"U" Originates from original "Unfired Pressure Vessel" Code (Section VIII)

"U-2" Pressure Vessels, Alternate Rules (Section VIII, Division 2)

"UV3" Pressure Vessels, High-Pressure Vessels (Section VIII, Division 3)

"U" Stamp

Requires the manufacturer to have a contract or agreement with a qualified inspection agency employing authorized inspectors

Certificate of authorization (COA) for using the "U" symbol is renewable every 3 years

"UM" Stamp

Exempt from inspection by authorized inspectors
Manufacturer is responsible for design, fabrication, inspection, and testing of the vessel
Limited in size and design pressure, cannot exceed
5 ft^3 volume and 250 psi, or
1.5 ft^3 volume and 600 psi
COA for using the "UM" symbol must be renewed annually
Some jurisdictions do not recognize the "UM" stamp and require inspection by an authorized inspector
Certificates of authorization (COA)
Allows manufacturers to bid and manufacturer ASME code vessels
Obligates manufacturers to quality of fabrication and documentation
Based on periodic surveys of the shop regarding its
Ability to design and fabricate pressure vessels
Quality control system

National Board of Boiler and Pressure Vessel Inspectors

Responsible for enforcing and administering the ASME code
Acceptance of a vessel by an inspector holding a National Board Commission signifies that the vessel complies with ASME code
Issues COA for the use of the "R" symbol stamp (repair)
Carries out certification of safety valves
A "V" symbol is placed on code-stamped relief devices

▶ ORGANIZATION OF THE ANSI/ASME B31 CODE FOR PRESSURE PIPING

Consists of a number of individually published sections
Most widely used design rules for pressure piping

The rules contained in each section reflect the kinds of piping installations that the reasonable subcommittee had in mind during the development of that section.

The kind of piping system to which a section is intended to apply is found in the beginning of each section under the heading "Scope"

Following are abbreviated scopes of all Sections to assist the owner in choosing the appropriate section

B31.1: Power Piping

Piping typically found in electric power generating stations, industrial and institutional plants, geothermal heating systems, and central and district heating systems (related to Section I)

B31.1 is intended to apply to the following:

- Piping for steam, water, oil, gas, air, and other services
- Metallic and nonmetallic piping
- All pressures
- All temperatures above −20°F (−29°C)
- Mandatory for piping that is attached directly to an ASME Section 1 boiler up to the first isolation valve, except in the case of multiple boiler installations where it is mandatory up to the second isolation valve

B31.3: Process Piping

Piping typically found in process facilities such as petroleum refineries, chemical, pharmaceutical, textile, paper, semiconductor, and cryogenic plants, and related processing plants and terminals

B31.3 is intended to apply to the following:

- Piping for all fluid services
- Metallic and nonmetallic piping

All pressures
All temperatures
The owner is responsible for determining designing when certain fluid services—Category M (toxic, high purity, high pressure, elevated temperature) or Category D (nonflammable, nontoxic fluids at low pressure and temperature)—are applicable to specific systems and for designing if a quality system is to be imposed

B31.4: Pipeline Transportation Systems for Liquid Hydrocarbons and Other Liquids

Piping transporting products that are predominately liquid among wells, plants, and terminals, and within terminals, pumping, regulating, and metering stations.
B31.4 is intended to apply to the following:
Piping transporting liquids such as crude oil, condensate, natural gasoline, natural gas liquids, liquefied petroleum gas, carbon dioxide, liquid alcohol, liquid anhydrous ammonia, and liquid petroleum products
Piping at pipeline terminals (marine, rail, and truck), tank farms, pump stations, pressure reducing stations, and metering stations, including scraper traps, strainers, and prover loops
All pressures
Temperatures from –20 to 250°F (–29 to 121°C) inclusive
B31.4 covers the design, construction, operation, and maintenance of these piping systems
B31.4 does not have requirements for auxiliary piping, such as water, air, steam, and lubricating oil

B31.5: Refrigeration Piping and Heat Transfer Components

Piping and heat transfer components containing refrigerants and secondary coolants including water when water is used as a secondary coolant

B31.5 is intended to apply to the following:

- Refrigerant and secondary coolant piping
- Heat transfer components such as condensers and evaporators
- All pressures and temperatures at or above −320°F (−196°C)

B31.8: Gas Transmission and Distribution Piping Systems

Piping transporting products that are predominately natural gas between sources and end-use services

B31.8 is intended to apply to the following:

- Onshore and offshore pipeline facilities used for the transport of gas
- Gas distribution systems
- Piping at compressor, regulating, and metering stations
- All pressures
- Temperatures from −20 to 450°F (−29 to 232°C) inclusive
- All two-phase flow lines from the well to the first process component

B31.8 covers the design, construction, operation, and maintenance of these piping systems, but it does not have requirements for auxiliary piping, such as water, air, steam, or lubricating oil

B31.9: Building Services Piping

Piping typically found in industrial, institutional, commercial, and public buildings and in multiunit residences

B31.9 is intended to apply to the following: piping for water and antifreeze solutions for heating and cooling, steam and steam condensate, air, combustible liquids, and other nontoxic, nonflammable fluids contained in piping not exceeding the following:

Dimensional limits

Carbon steel: NPS 42 (DN 1050) and 0.5 in. (12.7 mm) wall

Stainless steel: NPS 24 (DN 600) and 0.5 in. (12,7 mm) wall

Aluminum: NPS 12 (DN 300)

Brass and copper: NPS 12 (DN 300), 12.125 in. (308 mm) for copper tube

Thermoplastics: NPS 24 (DN 600)

Ductile iron: NPS 24 (DN 600)

Reinforced thermosetting resin: NPS 24 (DN 600)

Pressure and temperature limits, inclusive

Compressed air, steam, and steam condensate to 150 psi (1035 kPa gauge)

Steam and steam condensate from ambient to 366°F (186°C)

Other gases from ambient to 0 to 200°F (−18 to 93°C)

Liquids to 350 psi (2415 kPa) gauge and from 0 to 250°F (−18 to 121°C)

Vacuum to 14.7 psi (1 Bar)

Piping connected directly to ASME Section IV: Heating Boilers

B31.11: Slurry Transportation Piping Systems

Piping transporting aqueous slurries between plants and terminals and within terminals

B31.11 is intended to apply to the following:
- Piping transporting aqueous slurries of nonhazardous materials
- Piping in pumping and regulating stations
- All pressures and temperatures −20 to 250°F (−29 to 121°C)

B31.11 does not have requirements for auxiliary piping such as water, air, steam, lubricating oil, gas, and fuel

B31.12: Hydrogen Piping and Pipelines

Applicable to piping in gaseous and liquid hydrogen service and to pipelines in gaseous service

Applicable up to the joint connected to the pressure vessel or equipment

ANSI/ASME B31 Piping Code Committee

Prepares and issues new editions and addenda with dates that correspond with ASME code and addenda

Issue dates and mandatory dates do not always correspond with each other

▶ SOME OTHER PRESSURE VESSEL CODES AND STANDARDS IN THE UNITED STATES

Other codes and standards commonly used for the design of process vessels are listed here:

ANSI/API Standard 620: "Recommended Rules for Design and Construction of Large, Welded, Low-Pressure Storage Tanks," American Petroleum Institute (API), Washington, D.C.

ANSI/API Standard 650: "Welded Steel Tanks for Oil Storage," API, Washington, D.C.

ANSI-AWWA Standard D100: "Welded Steel Tanks for Water Storage," American Water Works Association (AWWA), Denver, Colorado

ANSI/AWWA Standard D101: "Inspecting and Repairing Steel Water Tanks, Standpipes, Reservoirs, and Elevated Tanks, for Water Storage," AWWA, Denver, Colorado

Standards of Tubular Exchanger Manufacturers Association, 6th ed., Tubular Exchanger Manufacture's Association, New York

Standards of the Expansion Joint Manufacturers Association, 4th ed., Expansion Joint Manufacturer's Association, New York

▶ WORLDWIDE PRESSURE VESSEL CODES

Other pressure vessel codes have been legally adopted in various countries

Difficulties occur in the following circumstances:

- Vessels are designed in one country
- Built in another country
- Installed in still another country

A partial summary of some of the various codes used in different countries follows.

Australia

Australian Code for Boilers and Pressure Vessels, SAA Boiler Code (Series AS 1200): AS 1210, Unfired Pressure Vessels and Class 1 H, Pressure Vessels of Advanced Design and Construction, Standard Association of Australia

Belgium

Code for Good Practice for the Construction of Pressure Vessels, Belgium Standard Institute (IBN), Brussels, Belgium

China

Regulations for the Design of Steel Pressure Vessels for Use in the Petroleum and Chemical Industries, Ministries of Petroleum Industry, Chemical Industry, and Machinery (and Ack. of M. of LaCour), Beijing, People's Republic of China

France

French Code of the Manufacturer of Unfired Pressure Vessels, Syndicate National de la Chaudreunerie et de la Tuyau Terie Industrielle (SNCT)

Germany

A.D. Merblatter Code, Carl Heymanns Verlag K.G., Kolin/ Berlin, Federal Republic of Germany

Italy

Italian Pressure Vessel Code, National Association for Combustion Control (ANCC), Milan, Italy

Japan

Japanese Pressure Vessel Code, Ministry of Labor, published by Japan Boiler Association, Tokyo, Japan

Japanese Standard, Construction of Pressure Vessels, JIS B 8243, published by the Japan Standards Association, Tokyo, Japan

Japanese High-Pressure Gas Control Law, Ministry of International Trade and Industry, published by the Institution for Safety of High-Pressure Gas Engineering, Tokyo, Japan

Netherlands

Rules for Pressure Vessels, Dienst Voor het Stoomwezen, The Hague, the Netherlands

Sweden

Swedish Pressure Vessel Code, Tryckkarls Kommission, the Swedish Pressure Vessel Commission, Stockholm, Sweden

United Kingdom

British Code B.S. 5500, British Standards Institution, London, England.

▸ ASME CODE, SECTION VIII, DIVISION 1 VERSUS DIVISION 2

ASME Code, Section VIII, is broken into Divisions 1, 2, and 3

Division 1

Most frequently used code in the world

Provides most economic design and construction for the majority of vessels in the petroleum industry

Uses low maximum allowable design stresses and implied design methods referred to as "design by rules"

Division 2

Developed to take advantage of technological advancements made in design methods, materials, and NDE

Higher design stresses are permitted because

- Combined stresses under all operating loads must be calculated
- Materials of construction must be tested more extensively
- Nondestructive examination of welded joints is more thorough

For same design conditions, requires thinner walls

Can be more economical for high-pressure heavy-wall vessels greater than 2 inches thick, when the reduction in material and fabrication costs exceeds the increase in design and inspection costs

▸ DESIGN CRITERIA, ASME CODE, SECTION VIII, DIVISION 1

Scope

Pressure vessels are defined as "Containers for the containment of pressure, either internal or external. This pressure may be obtained from an external source, or by the application of heat from a direct or indirect source, or any combination thereof."

Excluded from the scope:

- Piping and piping components
- Fired process tubular heaters
- Pressure containers that are integral parts or components of rotating or reciprocating mechanical devices (pumps, compressors, turbines, generators, engines, and hydraulic or pneumatic cylinders)

Also excluded are the following:

- Vessels of any size with an operating pressure of less than 15 psi

Vessels with an inside diameter, width, height, or cross-section diagonal under 6 inches

For pressures over 3000 psi

Additions and deviations from code rules may be necessary

If vessel complies with all requirements of Division 1, it may be stamped with the applicable code stamp

ASME code jurisdiction terminates at one of the following piping connections and depends on how the vessel was shipped to the owner/user:

First circumferential welded joint in welded-end connections

Face of the first flange in bolted flange connections

First threaded joint

Structure

Consists of three subsections plus mandatory and non-mandatory appendices.

Subsection A consists of Part UG, covering general requirements applicable to all pressure vessels

Subsection B covers specific requirements related to the various methods used in the fabrication of pressure vessels

It consists of the following:

Part UW: Welded Fabrication

Part UF: Forged Fabrication

Part UB: Brazed Fabrication

Subsection C covers specific requirements applicable to classes of materials used in pressure vessel construction

It consists of the following:

Part UCS: Carbon and Low-Alloy Steels

Part UNF: Nonferrous Materials
Part UHA: High-Alloy Steels
Part UCI: Cast Iron
Part UCL: Integral Cladding, Weld Metal Overlay Cladding, or Applied Linings
Part UCD: Cast Ductile Iron
Part UHT: Ferric Steels with Tensile Properties Enhanced by Heat Treatment
Part ULW: Layered Construction
Part ULT: Materials Having Higher Allowable Stresses at Low Temperature

Mandatory Appendices

Addresses specific subjects not covered elsewhere
Requirements are mandatory

Nonmandatory Appendices

Provide information and recommended practices

Allowable Stresses

Based on the maximum stress theory of failure
Most conservative method available
Only the primary membrane and bending stresses need to be considered for design
Detailed analysis of discontinuity, thermal, and fatigue stresses
Are not necessary
No design stress limits are imposed for them
Simplified approach is compensated for by using the following:

A safety factor of 3.5, which means the allowable tensile stress, at temperature, is approximately one fourth of the specified minimum tensile strength for the material
Special code-approved design rules, such as for the cone-to-cylinder junction
Code allowable stresses and used for the design of the important nonpressure parts, such as
Support skirts
Supports for important vessel internals
Allowable stresses for less-important nonpressure structural parts may be higher than code stresses
Code provides the necessary equations for the calculation of required thicknesses of vessel components resulting from internal and external pressure
Code requirements represent only minimum construction standards
For more complicated vessels or more severe services, higher-quality requirements are necessary
Requirements are usually defined in specifications written for typical vessels

Material Specifications

Materials for code pressure vessels must be made from code-approved material specifications

Section II, Part D

Provides tables with maximum allowable stress in tension at different temperatures for all the various materials of construction
Values should be used to determine the thickness of various pressure vessel parts using code equations and calculation procedures

When allowable stresses are requested for a new material not listed in the code, evaluation should be based on an evaluation of test data for this material

Appendix "B" (Nonmandatory), Section VIII, Division 1, lists the requirements to obtain approval of a new material

Following is a general discussion of common vessel materials approved by the code

Ferrous Alloys

Alloys having more than 50% iron

Includes the following:

- Carbon and low-alloy steels (less than 2% carbon)
- Stainless steels
- Cast iron (greater than 2% carbon)
- Wrought iron
- Quenched and tempered steels

ASME designates ferrous alloys with the prefix "SA"

Weldable carbon steels used in pressure vessels applications are limited to 0.35% carbon maximum; higher carbon content will make the steel very brittle and difficult to weld

Code Categories of Steels

Carbon steels

Low-alloy steels

High-alloy steels

Carbon Steels

Offer the proper combination of strengths, ductility, toughness, and weldability

Carbon content rarely over 0.25% (above this level, toughness and weldability are reduced) or below 0.15% for reasons of strength

Most significant alloying element is manganese, which is added to increase yield and tensile strength without reducing ductility

Limited to temperatures under 1000°F

Most widely used carbon steels are SA 36, SA 285 Grade C, and SA 515 and 516 Grade 70

Low-Alloy Steels

Main alloying elements are chromium (up to 10%) molybdenum, and nickel

Not as easily welded as carbon steel

More expensive than carbon steels, but their superior strength and toughness make them attractive

Carbon steel would require much heavier plates for the same load-carrying capacity

For high-temperature applications

In hydrogen service

Most common low-alloy steels used are SA 387 Grades 11 and 22

High-Alloy Steels

Usually called stainless steels

Known for high levels of corrosion resistance

Divided into three groups:

- Austenitic stainless steels
- Ferritic stainless steels
- Martensitic stainless steels

Steels from these three groups have the following characteristics:

- Austenitic stainless steels

Consist of chromium-nickel (300 series) and chromium-nickel-manganese (200 series)
Nonmagnetic, high corrosive resistant, and hardenable only by cold working (strength and hardness can be increased at the expense of ductility)
Only steels assigned allowable stresses in the code for temperatures above 1200°F
Most commonly used: 304 and 316; for welded construction: low-carbon Grades 304L and 316L or stabilized Grades 321 and 347
Higher chromium content steels, Grades 309 and 310, are resistant to oxidation and sulfur attack up to 2000°F

Ferritic stainless steels
Chromium stainless steels with a minimum of 10% chromium
Nonhardenable by heat treatment
Seldom used in vessel construction except for corrosion-resistant lining and cladding
Internal trays for less-corrosive environments
Common grades are Grades 405 and 430

Martensitic stainless steels
Contain 11% to 16% chromium with sufficient carbon to be hardenable (less than 1%)
Hardenable by heat treatment
Least corrosion resistant of the stainless steels
Difficult to form and require heat treatment after welding
Rarely used as construction materials for pressure vessel parts, with the exception of internal linings and trays, because of poor weldability

Nonferrous Alloys

Designated in the code by the prefix "SB"

Used in corrosive environments or at high temperatures where ferrous alloys are unsuitable

Broken into five categories

- Aluminum alloys
- Copper and copper alloys
- Nickel and high-nickel alloys
- Titanium and titanium alloys
- Zirconium

Aluminum Alloys

Excellent corrosion resistance

Lightweight and easy to form

Offer good strength-to-weight ratio

Major alloying elements are silicon, magnesium, and copper

Used for storage tanks or cryogenic vessels

Cooper-containing alloys (7xxx and 2xxx series alloys) may not have good corrosion resistance, depending on the service

Copper and Copper Alloys

Offer good corrosion resistance and machinability

Not susceptible to heat treatment

Strength can only be altered by cold working

Nickel and High-Nickel Alloys

Offer excellent corrosion resistance

Used in high-temperature applications in corrosive environments

Alloying elements are chromium (main), molybdenum, tungsten, aluminum, columbium, and titanium
Usually referred to by their commercial names rather than by their ASME designation numbers
Monel Inconel Incoloy Hastelloy

Titanium and Titanium Alloys
Used in vessels subject to severe environments
Offer high strength-to-weight ratios
High melting points
Excellent corrosion resistance
Can be hot-worked by standard steel mill equipment
Welding can be done only with helium or argon shielding

Zirconium
Has similar properties and uses to titanium

Inspection Openings
Regular internal inspections are very important to the safety and reliability of a vessel
Code Paragraph UG-40 requires inspection openings in all vessels (allows surfaces to be examined for defects):
Subject to internal corrosion, or
Having internal surfaces subject to erosion or mechanical abrasion
Removable heads or cover plates may be used in place of inspection openings provided they are large enough to do the following:
Permit entry
View of the interior

Pressure Relief Devices

Code Paragraph UG-125 requires all pressure vessels to have a means of protection against overpressure

When the source of pressure is external to the vessel, the protective device can be installed elsewhere in the system provided it can control the pressure in such a manner that it will not exceed the MAWP at the operating temperature

Pressure relief devices do not have to be provided by the vessel manufacturer, but must be installed before the vessel is placed in service or tested for service

The number, size, and location of the pressure relief devices must be listed on the manufacturer's data report

Information on the design and application of pressure relief devices is included in another section

Cryogenic and Low-Temperature Vessels

Carbon steels and low-alloy steels exhibit a significant loss of toughness and ductility at cold temperatures

At low temperatures, metal becomes brittle, and without prior noticeable deformation, fracture can occur at stresses considerably below the yield point (brittle fracture)

Materials used for low-temperature applications must have a proved ability to safely resist brittle fracture

Code Paragraph UCS-66 states whether an impact test is required to verify toughness based on a combination of minimum design metal temperature and thickness

Impact Testing Requirements

When the calculated stress in tension for a design condition is less than the maximum allowable design stresses, the design temperature not requiring impact testing can be further lowered based on Code Figure UCS-66 "Impact Test Exemption Curves" (refer to Code Paragraph UCS-66(a))

If impact testing is required, Code Paragraph UG-84 describes how it must be done; this method is discussed in more detail in a later section

Code Paragraph UHA-51 sets impact test requirements for stainless steels

Code Paragraph UNF-65 covers nonferrous materials

Whenever impact tests are necessary, every part of the vessel, including welds, must comply with code requirements

Testing, Inspection, and Stamping

ASME code has three quality levels based on the type and degree of examination of the pressure vessel:

Visual examination of butt welds

Spot radiographic examination of butt welds

Full radiographic examination of butt welds

By applying joint efficiencies and stress multipliers, the code compensates for a lower level of quality with an increased safety factor

Theoretical factors of safety

Level (1): 5

Level (2): between 3.5 and 4

Level (3): 4

Code requires the following:

Radiography pressure vessels for certain design and service conditions; the type of radiography depends on the following characteristics:

Materials and thickness

Joint efficiency used in design calculations
Design of weld joints
Employing a radiographic technique that will reveal the size of defects having a dimension equal to or greater than 2% of the thickness of the base metal

Weld quality depends on the following:
Design of the joint
Welding procedures
Fill-up of the parts
Operator qualifications
Supervision
Examination of the weld

All methods of testing and inspecting welds can be grouped into two categories:
Destructive
Nondestructive

Destructive Tests

Use methods to determine the properties of a weld where the weld itself is usually destroyed

Used to determine a number of weld properties:
Ductility
Weld penetration
Tensile strength and fusion
Methods include bend test, impact test, tensile test, and etch test

Not considered a substitute for spot-radiographic examination and permit no increase in joint efficiency

Nondestructive Tests

Use methods to determine the suitability of a weld for the service conditions to which it will be subjected

Do not affect or alter the structure or appearance of the weld and thus are not as thorough as destructive tests in determining the properties of a weld

Nondestructive Examination (NDE)

Requires procedures that, like welding procedures, must do the following:
- Satisfy the requirements of the code
- Be acceptable to the authorized inspector

Various sections of the code refer to Section V: Nondestructive Examinations

Manufacturer quality control manuals must include the following:
- Description of the manufacturer's system for assuming that all NDE personnel are qualified in accordance with the applicable code
- Written procedures to assure that NDE requirements are maintained in accordance with the code
- If NDE is subcontracted, a statement of who is responsible for checking the subcontractor's site, equipment, and qualification of procedures is required to assure they are in compliance with the applicable provisions of the code

Hydrostatic Testing

All completed pressure vessels must pass a hydrostatic test as prescribed in Code Paragraph UG-99

Test pressure must be at least 1.3 times the MAWP to be marked on the vessel (1.5 times for vessels built prior to July 99 Addenda)

Vessel and its contents are recommended to be at least 30°F above the minimum design metal temperature (MDMT) to prevent brittle fracture during testing

After the test, all joints and connections should be inspected at a pressure not less than two thirds of the test pressure

Manufacturer's quality manual must contain a description of the system used for assuring that the gauge (minimum of two) measuring and testing devices are properly calibrated (minimum every 6 months) and controlled

Code Paragraph UG-102 establishes requirements for the use of test gauges

Pneumatic Testing

Code Paragraph UG-100 allows for pneumatic testing where vessels cannot be safely filled with water

Testing is limited to 1.1 times MAWP (1.25 for vessels built prior July 1999 Addenda)

Pneumatic tests should be avoided whenever possible; the energy stored in a pressure vessel is much greater with a compressible gas

Inspection by an "Authorized Inspector"

Reviews design documents submitted by manufacturer

Verifies that all materials to be used comply with the mill test reports and code specifications

Dimensions and thicknesses must meet code tolerances and a thorough examination is made for defects, such as lamination, pit marks, and surface scars

For the whole fabrication process, the inspector will indicate "hold points" and discuss with the manager of quality control arrangements for additional inspections, including the review of radiographs

After witnessing the final hydrostatic or pneumatic test and carrying out the necessary examinations, the inspector will do the following:

Check the vessel data report sheets to make sure they are properly filled out and signed by the manufacturer

Check that the material and quality of work in the vessel are accurately recorded

Then the inspector will allow the code stamp to be applied

Code Stamp

When construction is completed in the field, some organization must take responsibility for the entire vessel:

Issuing of data sheets on the parts fabricated in the shop and in the field

Application of the code stamp

Usually the vessel manufacturer but may be a separate assembler provided the assembler has the following:

Required ASME certification of authorization

Code symbol stamp

Stamp can be applied directly on the vessel but is usually applied to a nameplate

Nameplate should be located in a conspicuous place preferably next to a manhole or handhole opening

Code Paragraph UG-116 states what information should be marked on each code-built pressure vessel

▶ DESIGN CRITERIA, ASME CODE, SECTION VIII, DIVISION 2

Scope

Pressure vessels subject to direct firing, but not within the scope of Sections I, III, or IV, can be built in accordance with Division 2

Requirements of Section VIII, Division 1, have been upgraded

A detailed stress analysis of the vessel, referred to as "design by analysis," is required

Many restrictions in the choice of materials and some common design details are prohibited

More thorough materials certification and NDE are required because design stresses are higher than those allowed in Division 1

All these restrictions allow the use of a theoretical design factor of safety of 3, compares to 3.5 for Division 1

There is no pressure limitation, but at very high pressures, additions and deviations from code rules may be necessary in order to meet design and construction requirements

Structure

Consists of eight parts plus mandatory and nonmandatory appendices

Part A6: General Requirements

Covers the scope of the division, establishes its jurisdiction, and states the responsibilities of the user and manufacturer and the duties of the inspector

Part AM: Material Requirements

Lists the materials that may be utilized, special requirements, and information concerning materials properties

Part AD: Design Requirements

Part AF: Fabrication Requirements

Part AI: Inspection and Radiography

Part AT: Testing

Part AS: Marking, Stamping, Reports, and Records

Allowable Stresses

Division 2 employs the maximum shear stress theory of failure

More accurate for predicting the failure of a pressure vessel by ductile yielding than the maximum stress theory utilized by Division 1

Assumes yielding of a vessel at any location is determined by the maximum shear stress resulting from the combination by all of the loads acting upon the vessel at the location

All categories of stress (primary, secondary, and peak) must be added together to determine the stress intensity when designing a component, but the maximum stress intensity permitted for design depends on the particular categories of stress

Code Appendix 4 provides the details for determining the maximum stress intensity permitted for design with various combinations of stress in the different categories

In general, the criteria for determining maximum stress intensities for design are the following:

1. The primary membrane stress cannot exceed the design stress for the material of construction at the design temperature
2. Local primary membrane stresses are permitted to reach 1.5 times the design stress
3. Primary membrane plus primary bending stresses are permitted to reach 1.5 times the design stress
4. Primary membrane plus secondary and peak stresses are permitted to reach 3 times the design stress
5. An additional 20% increase in stress is permitted for 1, 2, and 3 above for adding intermittent loads such as wind and earthquake loads

Part AD provides simplified "rules" for the design of most of the commonly used components

Consistent with the "design by analysis" concept that incorporates the maximum shear stress theory

Actual design procedure is not much more complicated than that required by Division 1

- Code Appendix 4 details the stress analysis procedures that must be used for components that do not conform to the simplified rules in Part AD
- It is the responsibility of the designer to determine the following:
  - When the simplified rules of Part AD can be used
  - When a stress analysis conforming to Appendix 4 must be used
- Code Appendix 5 (mandatory), "Design Based on Fatigue Analysis"
  - Provides the method of design for cyclic loading
- Code Figure 5-100.1 presents fatigue curves for the type of materials permitted by Division 2, which gives an allowable number of cycles for certain values of stress intensities
- Two different loading cases are considered, depending on whether the directions of the principle stress changes
- Code Appendix 6 (mandatory), "Experimental Stress Analysis"
  - Provides the procedure for tests to be run in order to determine governing stresses, collapse load, or the adequacy of a component for cyclic loading
  - Should be used if theoretical stress analysis for a part of a pressure vessel proves to be inadequate

Materials and Tests

- More restrictive than Division 1 in the choice of materials that can be used
- Special inspection and testing requirements
- Add to the reliability of the materials
- Permit the use of higher allowable stress values

Examples:

Plate and forgings over 4 inches thick require ultrasonic examination (Code Paragraph AM-203)

Cut edges of base materials over 1½ inches thick must be inspected for defects by liquid penetrant or magnetic particle methods

Because of the higher stress allowed and for protection against brittle fracture of some steels like SA 285, thicknesses over 1 inch are limited to 65°F or above, unless they are impart tested (Code Paragraph AM-211)

In Division 1, those commonly used plate steels allow temperature as low as −20°F without an impact test

Part AF, Article F-2, allows only some welding processes, and special precautions have to be taken before and during welding

Division 2 provided for only one quality vessel

All longitudinal and circumferential seams butt welded and 100% radiographic

Requires more NDE than Division 1

Radiographers and ultrasonic examiners

Must be qualified (Code Paragraph AI-501 and Code Appendix 9-3) and qualifications documented as required

Hydrostatic Testing

Unlike Division 1, establishes an upper limit on the test pressure

No less than 1.25 times the design pressure at every point in the vessel

If at any point this pressure (including static head) is exceeded by more than 6%, the upper limit shall be established so that the calculated stress intensities, using all the loadings that may exist during the test, do not exceed certain limits given in Paragraph AD-151.1

Pneumatic Testing

Minimum test pressure is 1.15 times the design pressure

Responsibilities

The user of the vessel or a representative agent must provide the manufacturer with design specifications giving detailed information about the intended operating conditions (Code Paragraph AG-301)

Information should constitute basis for the following:

Selecting materials

Designing, fabricating, and inspecting the vessel

User's Design Specification

Should include the method of supporting the vessel

List the amount of corrosion allowance to provide

Information on whether or not a fatigue analysis of the vessel should be made for cyclic service, and if it is required, to provide enough information so that an analysis can be carried out

Must be certified by a registered professional engineer experienced in pressure vessel design

Manufacturer's Design Report

Prepared by vessel manufacturer

Establishes conformance with the rules of Division 2 for the design conditions specified in the user's design specification

Conformance should also be certified by a registered professional engineer

Authorized Inspector's Verification

Not responsible for approving design calculations

Verifies existence of user's design specification and manufacturer's design reports and makes sure that they have been certified by a registered professional engineer (Code Paragraph AG-303)

Reviews the design with respect to materials used; to see that the manufacturer has qualified welding procedures and welders; and to review the vessel geometry, weld details, nozzle attachment details, and nondestructive testing requirements

► ASME CODE, SECTION IX: WELDING

ASME Code Requirements

Does not dictate welding methods to the manufacturer

Does require proof that the welding is sound (Code Paragraph QW-201)

Does require that welders are qualified in accordance with ASME Code, Section IX, "Welding and Performance Qualifications" (Code Paragraph QW-301)

Additional information on welding is included in another section

Welding Procedure Specifications (WPS) and Procedure Qualification Records (PQR)

Required by Section IX

Most important factors in the achievement of sound welding

If properly established, welders need only to demonstrate their ability to make a weld that will pass root and face or side bend tests in the position to be welded, in order to gain qualification

Post-Weld Heat Treatment (PWHT)

Required on some pressure vessels

Dependent on materials of construction and shell thickness

Code Paragraph UW-40 describes the procedure to be followed for PWHT, when required

Must be done before hydrostatic testing and after any welded repairs

Manufacturer must describe in its quality control manual

Welding preheat and interpass temperature control methods

For Post Weld Heat Treatment (PWHT) the manufacturer must describe the following:

- How the furnace is loaded
- How and where the thermocouples are placed
- Time-temperature recording system
- Methods used for heating and cooling
- Metal temperatures and how they are controlled
- Methods for supporting the vessel during PWHT

ASME CODE, SECTION I, POWER BOILERS

Power boilers are not in the scope of this course

Provisions of Section I are briefly discussed for comparison purposes only

Scope and Limitations

Covers rules for construction of the following items:

Power boilers

Electric boilers

Miniature boilers

High-temperature water boilers used in stationary service

Power boilers used in locomotive, traction, and portable service

Applies to the boiler and its external piping

Superheaters, economizers, and other pressure parts connected directly to the boiler without intervening valves are considered part of the boiler and thus are included within the scope

Includes fired steam boilers (fired pressure vessels) where steam is generated by the application of heat resulting from the combustion of fuel

Includes unfired pressure vessels in which steam is generated, with the following exceptions:
- Evaporators
- Heat exchangers
- Vessels in which vapor is generated incidental to the operation of a processing system, containing a number of pressure vessels, such as used in the chemical and petroleum manufacture

Boilers are those in which steam or other vapor is generated at a pressure of more than 15 psi

High-temperature water boilers are intended for operation at pressures exceeding 160 psig or temperatures exceeding 250°F

For vessels operating below these levels, Section IV "Heating Boilers" apply

If all applicable requirements of Section I are met, these boilers can still be stamped in accordance with Section I

Allowable Stresses

Materials and allowable design stresses are the same as for Section VIII, Division 1

Not all pressure vessel materials are suitable for the construction of boilers; for example, nonferrous materials are limited to copper and nickel alloys

Section I is less definitive than Section VIII, Division 1, concerning the following:
- Loadings necessary to consider
- What shall be included in a design specification or purchase order

Code Paragraph PG-22 states that loadings that cause stresses to go higher than 10% over those caused by internal design pressure and static head should be considered

A safety factor of 3.5 is used for determining the allowable stresses

Fabrication, Inspection, and Tests

Requires all welds in pressure parts to be
- Full penetration butt welds
- Fully radiographed (except for pipe materials under 10 inches and under nominal pipe size or 1⅛-inch wall thickness having only circumferential weld butt joints)
- Weld procedures and welders must be qualified according to Section IX
- Carbon steel parts over ¾-inch thick require PWHT

Inspection by an authorized inspector is required

A hydrostatic test is required at a pressure not less than 1.5 times the MAWP, but this test pressure must not exceed 90% of a material's specified yield strength

Provides for the following six code symbol stamps:
- "S": Power boilers
- "M": Miniature boilers
- "E": Electric boilers
- "A": Boiler assemblies
- "PP": Pressure piping
- "V": Safety valves

ADDITIONAL REQUIREMENTS EMPLOYED BY USERS IN CRITICAL SERVICE

ASME codes are the minimum requirements for the construction of pressure vessels

Higher requirements are normally employed by users in process facilities and other critical services

Additional Requirements

Restrict the use of some design details permitted by the code (especially for nozzles and weld joints)

Specify more materials testing to assure adequate toughness to restrict brittle fracture

Specify more thorough inspection to ensure the integrity of fabrication

Significant requirements above the code minimum requirements include the following:

- Nozzle welds are normally recommended to be full penetration welds that receive either ultrasonic or magnetic particle inspection
- Flange welds are normally recommended to be full penetration welds
- Integral reinforced nozzles are recommended for shell components over 2 inches thick, design temperatures over 650°F, and for low-temperature service when Charpy V Notch-impact testing is required
- Longitudinal and girth welds should be double-V butt welds or an acceptable equivalent
- Hydrotest pressure should stress all longitudinal welds to at least 1.5 times the design pressure, and the design pressure cannot be limited by flanges, nozzles, or reinforcing pads
- PWHT is recommended for some process environments and may exceed ASME code requirements

2

Materials of Construction

▶ OVERVIEW

This chapter discusses the following topics:

Material selection
Nonferrous alloys
Ferrous alloys
Heat treatment of steels
Brittle fracture
Hydrogen embrittlement

▶ MATERIAL SELECTION

General Considerations

The majority of pressure vessels are constructed of ferrous and nonferrous alloys

Ferrous Alloys

Defined as having more than 50% iron
Used in ASME Code, Section VIII, Divisions 1 and 2
Include carbon and low-alloy steels, stainless steels, cast iron, wrought iron, and quenched and tempered steels
ASTM designates all ferrous alloys by the letter "A"
ASME uses "SA"

Nonferrous Alloys

Include aluminum, copper, nickel, titanium, and zirconium
ASTM designates nonferrous alloys by the letter "B"
ASME uses "SB"
In most cases, ASME and ASTM material specifications are identical

Vessels built to the ASME code usually refer to ASME specifications

Selection of materials that are adequate for a given process is complicated and depends on factors such as corrosion, strength, and cost

Corrosion

Defined as the deterioration of metals by chemical action

Single most important consideration in selecting materials

A slight change in the chemical composition of the environment can significantly change the corrosive behavior of a metal

In highly corrosive environments, every phase of the pressure vessel fabrication process, such as burning, forming, welding, stress relieving, and polishing, must be evaluated for corrosion

The cleanliness and finish of the inside surface of a pressure vessel before it is placed in operation are very important for preventing subsequent corrosion in service

Strength

The strength level of a material has a significant influence on its selection for a given application; this is especially true at elevated temperatures where

- Yield and ultimate strength are relatively low
- Creep and rupture behavior may control the allowable stress values

ASME Code, Section VIII, Division 1 criteria for allowable stress at elevated temperatures take into account both

Creep

Rupture behavior

In applying the ASME criteria for allowable stress, the following procedures are used:

Specified minimum yield stress

Specified minimum tensile stress

Creep rate

Rupture strength

Specified Minimum Yield Stress

Test data are plotted at various temperatures (Figure 2.1)

Trend curve is drawn through the averages of the data for individual test temperatures

The specified minimum-yield-stress curve is obtained by multiplying the yield trend curve values by the ratio of the specified minimum value, as given in the material specification, to the trend value at room temperature.

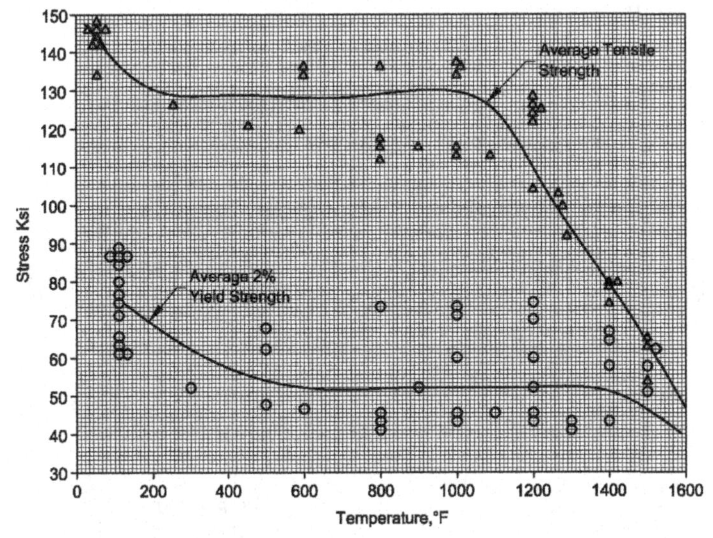

Figure 2.1 Tensile and yield strength.

Specified Minimum Tensile Stress

Tensile trend curve is determined by the same method as the yield trend curve including the ratio factor

The specified minimum tensile stress is arbitrarily taken as 110% of the tensile trend curve

Creep Rate

To establish the creep rate of 1%/100,000 hours, data are plotted as shown in Figure 2.2

Interpolation and extrapolation may be needed to establish the creep rate for various temperature levels

Rupture Strength

Test data are normally plotted as shown in Figure 2.3

In some cases, the data need to be extended to 100,000 hours so that extreme care must be taken to extrapolate accurately

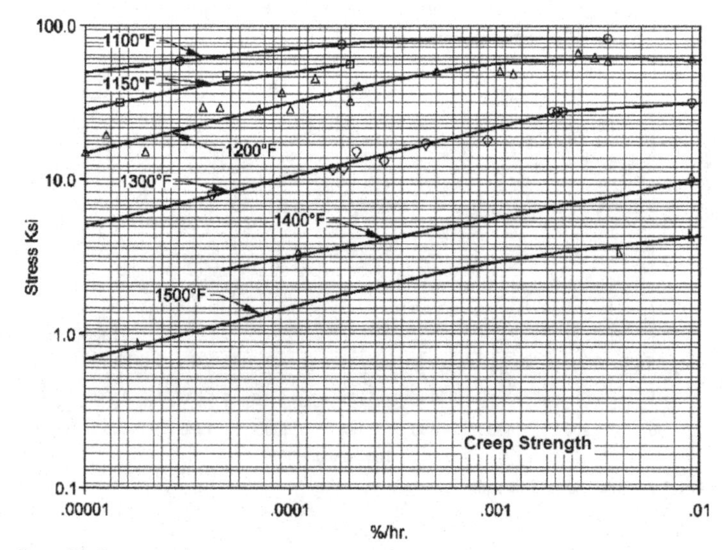

Figure 2.2 Creep strength.

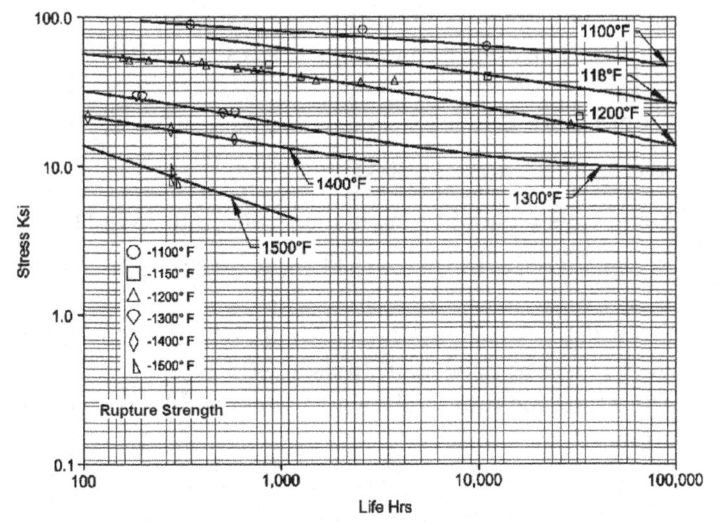

Figure 2.3 Rupture strength.

Example 1

A user is requesting code approval for a new material that has a minimum specified tensile stress of 120 ksi and a minimum specified yield stress of 60 ksi at room temperature. Tensile and yield values for various heats and temperatures are shown in Figure 2.1. Creep and rupture data are given in Figures 2.2 and 2.3, respectively. What are the allowable stress values at 300° and 1200°F based on ASME Code, Section VIII, Division 1 criteria?

Solution

Allowable stress at 300°F.

1. From Figure 2.1,

Average tensile = 130 ksi

Tensile stress reduced to minimum = (130)(120/140) = 111 ksi

Specified minimum tensile stress = (111)(1.10) = 12.2 ksi

Maximum stress to be used cannot exceed 120 ksi
Allowable stress based on tensile stress = (120/4) = 30 ksi

2. From Figure 2.1,

   Average yield stress = 60 ksi

   Yield stress reduced to minimum = (60)(60/75) = 48 ksi

   Allowable stress based on yield stress = (48)(2/3) = 32 ksi

3. From Figures 2.2 and 2.3, it is apparent that creep and rupture are not considerations at 300°F.
4. Therefore, maximum allowable stress at 300°F is 30 ksi.

Solution

Allowable stress at 1200°F.
From Figure 2.1,

Average tensile = 112 ksi

Tensile stress reduced to minimum = (112)(120/140) = 96 ksi

Specified minimum tensile stress = (96)(1.10) = 106 ksi, which is less than the maximum allowed (120 ksi)

Allowable stress based on tensile stress = (106/4) = 26.5 ksi

1. From Figure 2.1,

Average yield stress = 52 ksi

Yield stress reduced to minimum = (52)(60/75) = 42 ksi

Allowable stress based on yield stress = (42)(2/3) = 28 ksi

Code Section	Note	Minimum Specified Tensile Strength	Tensile Strength at Temperature[a]	Minimum Specified Yield Strength	Yield Strength at Temperature	Creep Rate of 0.01% in 1000 hrs	Stress to Rupture in 100,000 hrs	
						Average	Average	Minimum
Section I	1	$^{1}/_{4}$	$^{1}/_{4}$	$^{2}/_{3}$	$^{2}/_{3}$	1.00	0.67	0.80
	2	$^{1}/_{4}$	$^{1}/_{4}$	$^{2}/_{3}$	0.90	1.00	0.67	0.80
Section III,						---	---	---
Class 1	3	$^{1}/_{3}$	$^{1}/_{3}$	$^{2}/_{3}$	$^{2}/_{3}$	---	---	---
	4	$^{1}/_{3}$	$^{1}/_{3}$	$^{2}/_{3}$	0.90	---	---	---
Bolting	5	---	---	$^{1}/_{3}$	$^{1}/_{3}$	---	---	---
Classes 2 and 3	6	$^{1}/_{4}$	$^{1}/_{4}$	$^{2}/_{3}$	$^{2}/_{3}$	---	---	---
	7	$^{1}/_{4}$	$^{1}/_{4}$	$^{2}/_{3}$	0.90	---	---	---
Bolting	8	$^{1}/_{4}$	$^{1}/_{4}$	$^{2}/_{3}$	$^{2}/_{3}$	---	---	---
	9	$^{1}/_{5}$	$^{1}/_{4}$	$^{1}/_{4}$	$^{2}/_{3}$	---	---	---
Section IV	1	$^{1}/_{5}$	---	---	---	---	---	---
Section VIII,								
Division 1	10	$^{1}/_{4}$	$^{1}/_{4}$	$^{2}/_{3}$	$^{2}/_{3}$	1.00	0.67	0.80
	11	$^{1}/_{4}$	$^{1}/_{4}$	$^{2}/_{3}$	0.90	1.00	0.67	0.80

Figure 2.4 ASME code allowable tensile strength multiplying factors (based on the 1998 edition).

Bolting	8	1/4	1/4	2/3	2/3	1.00	0.67	0.80
	9	1/5	1/4	1/4	2/3	1.00	0.67	0.80
Division 2	10	1/3	1/3	2/3	2/3			
	11	1/3	1/3	2/3	0.90			
Bolting	12	1/4	1/4	2/3	2/3			
	13	1/5	1/4	1/4	2/3			
	14	---	---	1/3	1/3			

NOTES

1. Minimum for all materials.
2. For austenitic SS and nickel alloys only.
3. Minimum for all materials except bolting, austenitic SS, Ni-Cr-Fe, and Ni-Fe-Cr.
4. Minimum for austenitic SS, Ni-Cr-Fe, and Ni-Fe-Cr.
5. Minimum for bolting.
6. Minimum for ferritic steels.
7. Minimum for austenitic SS and nonferrous.
8. Minimum for all bolting except heat-treated.
9. Minimum for all heat-treated bolting.
10. Minimum for all materials except bolting.
11. Increase permitted for austenitic SS and nickel alloy.
12. Minimum for Appendix 3 of Section VIII-2 boiling except heat-treated.
13. Minimum for Appendix 3 of Section VIII-2 heat-treated bolting.
14. Minimum for Appendix 4 of Section VIII-2 bolting.

*Values in this column are multiplied by 1.1.

Figure 2.4 (*Continued*).

2. From Figure 2.2,

$$\text{Creep stress for } 0.01\% \text{ in } 1000 \text{ hours} = 15 \text{ ksi}$$

$$\text{Allowable stress based on creep} = 15 \text{ ksi}$$

3. From Figure 2.3 and 2.4,

$$\text{Stress to cause rupture at } 10^6 \text{ hours} = 22 \text{ ksi}$$

$$\text{Allowable stress based on rupture} = (0.67 \times 22) = 14.7 \text{ ksi}$$

4. Therefore, maximum allowable stress at 1200°F is 14.7 ksi

Material Cost

Because the costs of materials vary significantly, the designer should evaluate material cost versus other factors such as the following:

Corrosion
Expected life of equipment
Availability of material
Replacement cost
Code restriction on fabrication and repairs

A summary of the cost of some frequently used materials is given in the following table. In view of the large difference in cost, the designer should consider all factors carefully.

Approximate Cost of Materials Used in Pressure Vessel Construction

TYPE	COST (US $/POUND)
Carbon steel	1.50
Low-alloy steel	3.75
Stainless steel	4.50–12.50
Aluminum	7.50
Copper, bronze	8.75
Incoloy	20.00

(Continued)

TYPE	COST (US $/POUND)
Monel	25.00
Inconel	30.00
Hastelloys	75.00
Titanium	75.00
Zirconium	100.00
Tantalum	500.00

As of January 2012.

NONFERROUS ALLOYS

General Considerations

ASME Code, Section VIII, Division 1 lists the following five nonferrous alloys for code consideration:

Aluminum

Copper

Nickel

Titanium

Zirconium

These alloys are normally used in the following situations:

Corrosive environment

At elevated temperatures where ferrous alloys are unsuitable

Nonferrous alloys are nonmagnetic except for commercially pure nickel, which is slightly magnetic

Aluminum Alloys

Offer unique properties that make them useful in process equipment:

Nonmagnetic

Light in weight

Have good formability

Excellent weight-strength ratio

Aluminum surfaces exposed to the atmosphere form an invisible oxide skin that protects the metal from further oxidation and thus aluminum has a high resistance to corrosion

Aluminum alloys have a systematic numbering system as shown in Figure 2.5

Specification number also designates the various product forms (SB 209-plates and SB 210-drawn seamless tube)

First digit of the alloy designation number indicates its major alloying element as shown in the figure

Categorized by the following (refer to Figures 2.5 and 2.6):

ASME specification number

Alloy designation

Temper designation

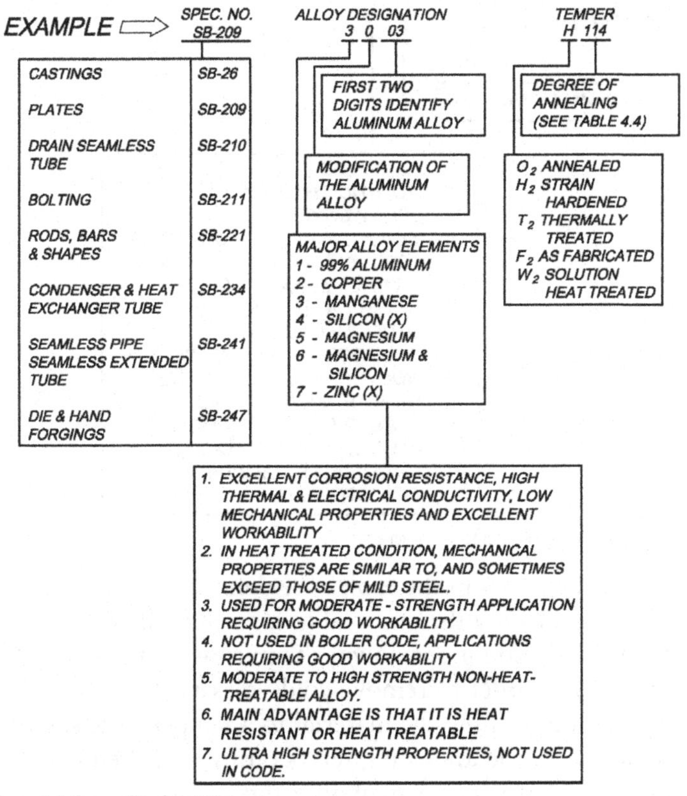

Figure 2.5 Copper alloy designation.

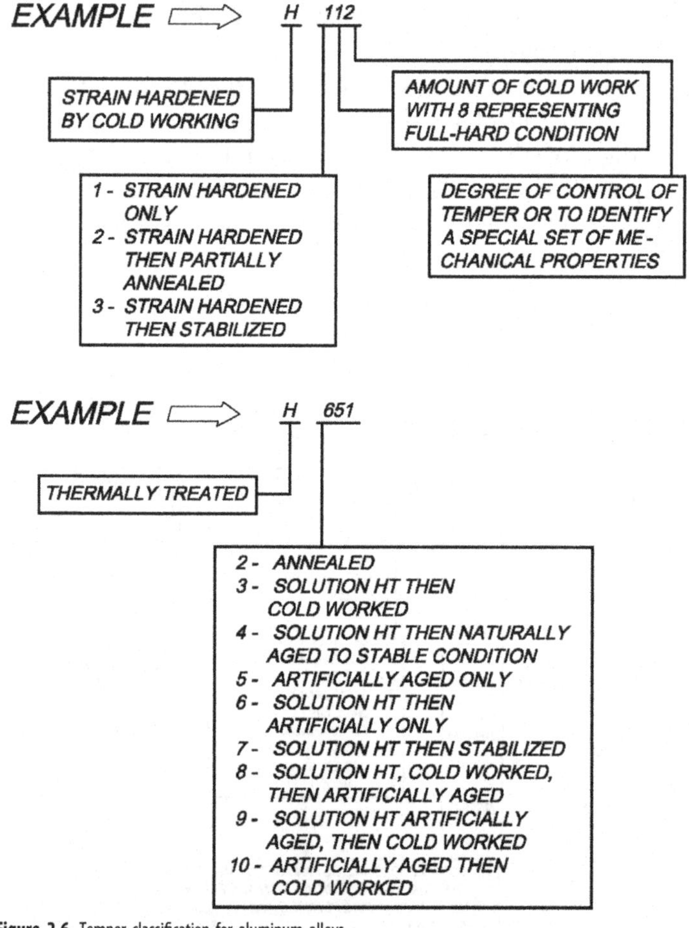

Figure 2.6 Temper classification for aluminum alloys.

Some of the terms in the tables are defined as follows:

Annealing: heating the material to a given temperature and then slowly cooling it down; the purpose is to soften the material in order to remove cold-working stress

Normalizing: heating the material to a temperature slightly higher than the annealing temperature and then cooling at a rate that is faster than annealing

Solution heat treating: heat treating at a temperature high enough for the alloys to be randomly dispersed

Stabilizing: low-temperature heating to stabilize the properties of the alloy

Strain hardening: modification of metal structure by cold working, resulting in an increase in strength with a loss in ductility

Thermal treating: temperature treatment of an alloy to produce a stable temper

Copper and Copper Alloys

Most copper alloys are used because of their good corrosion resistance and machinability

They are homogenous and thus not susceptible to heat treatment

Strength may be altered only by cold working

Alloy designation system serves to identify the type of material, as shown in the following table

Alloys 101 to 199 are high-grade copper with very few alloys added

Alloys 201 to 299 are brasses (mainly copper and zinc)

Alloys 501 to 665 are bronzes, composed of copper and elements other than zinc

Other properties of copper alloys are also shown

Most copper alloys are distinguishable by their color, except for C_u-N_i alloys, which tend to lose their color as the amount of N_i is increased

Nickel and High-Nickel Alloys

Have excellent corrosion and oxidation resistance

Ideal for high-temperature applications with corrosive environments

Alloy Designation of Coppers

Copper Alloys	
101-199	Coppers
201-299	Copper-zinc alloys (brass)
301-399	Copper-zinc alloys (leaded brass)
401-499	Copper-zinc-tin alloys (tin brass)
501-599	Copper-tin alloys (phosphor bronze)
601-645	Copper-aluminum alloys (aluminum bronze)
645-666	Copper-silicon alloys (silicon bronze)
666-699	Miscellaneous copper alloys
701-730	Copper-nickel alloys
Cold-Worked Temper Designations Description	**Approximate Percentage Reduction by Cold Working**
Quarter hard	10.9
Half hard	20.7
Three-quarters hard	29.4
Hard	37.1
Extra hard	50.0
Spring	60.5
Extra spring	68.7

Normally called by their commercial names rather than their ASME designation number, as shown on following table

Titanium and Zirconium Alloys

Used in process equipment subjected to severe environment

In ASME Code, Section VIM, Division 1

Unalloyed titanium is listed for Grades 1, 2, and 3

Alloyed titanium is listed for Grades 7 and 16

Unalloyed zirconium 702 and alloyed zirconium 705

Modules of elasticity and coefficient of thermal expansion of both is about half that of steel

Density of zirconium is slightly less than that of steel, whereas the density of titanium is about 0.58 times that of steel

Nonferrous Alloys

TRADE NAME (NOR-FERROUS ASME NUMBER)	COMPOSITION	ALLOY DESIGNATION	ASME NUMBER				
			PLATE AND SHEETS	PIPE AND TUBE	TUBE	RODS, BARS SHAPES, AND FORGINGS	BOLT
Nickel (02200)	Ni	200	SB-162	SB-161	SB-163	SB-160	SB-160
Nickel (02201)	Ni-low-C	201	SB-162	SB-161	SB-163	SB-160	SB-160
Monel (04400)	Ni-Cu	400	SB-127	SB-165	SB-163	(SB-164 SB-564)	SB-164
Monel (04405)	Ni-Cu	405	—	—	—	SB- 164	SB-164
Inconel 600 (06600)	Ni-Cr-Fe	600	SB-168	(SB-167 SB-517)	SB-163 SB-516	(SB-166 SB-564)	SB-166
Inconel 625 (06625)	Ni-Cr-Mo-Cb	625	SB-443	SB-444	—	SB-446	SB-446
Inconel 690 (06690)	Ni-Cr-Fe	690	SB-168	SB-167	SB-163	SB-166	SB-166
Incoloy 800 and 800H (08800 and 08810)	Ni-Fe-Cr	(800 and 800H)	SB-409	(SB-407 SB-514)	SB-163 SB-515	(SB-408 SB-564)	SB-408
Incoloy 825 (08825)	Ni-Fe-Cr-Mo-Cu	825	SB-424	SB-423	SB-163	SB-425	SB-425
Hast. B-2 (10665)	Ni-Mo	B-2	SB-333	SB-619	(SB-622 SB-626)	SB-335	SB-335
Hast. C-4 (06445)	Ni-Mo-Cr	C-4	SB-575	SB-619	(SB-622 SB-626)	SB-574	SB-574

(*Continued*)

Nonferrous Alloys—cont'd

TRADE NAME (NOR-FERROUS ASME NUMBER)	COMPOSITION	ALLOY DESIGNATION	ASME NUMBER				
			PLATE AND SHEETS	PIPE AND TUBE	TUBE	RODS, BARS SHAPES, AND FORGINGS	BOLT
Hast. C-276 (10276)	Ni-Mo-Cr	C-276	SB-575	SB-619	(SB-622 SB-626)	SB-574	SB-574
Hast. G (06007)	Ni-Cr-Fe-Mo-Cu	G	SB-582	SB-619	(SB-622 SB-626)	SB-581	SB-581
Hast. G-2 (06975)	Ni-Cr-Fe-Mo-Cu	G-2	SB-582	SB-619	(SB-622 SB-626)	SB-581	SB-581
Hast. G-3 (069857)	Ni-Cr-Fe-Mo-Cu	G-3	SR-582	(SB-622 SB-619)	SB-626	SB-581	—
Carp. 20 (08020)	Cr-Ni-Fe-Mo-Cu-Cb	20 Cb	SB-463	SB-464	SB-468	(SB-462 SB-473)	SB-462
Rolled alloy 330 (08330)	Ni-Fe-Cr-Si	330	SB-536	SB-535	SB-710	SB-511	—
904L (08904)	Ni-Fe-Cr-Mo-Cu-low-C	904L	SB-625	SB-677 SB-674	SB-677	SB-649	—

▶ FERROUS ALLOYS

General Considerations

Effect of carbon content in iron alloys
Cast iron: more than 2%
Steel: less than 2%
Hypereutectoid steels: greater than 0.8% and less than 2%
Hypoeutectoid steel: less than 0.8%
Most steels used in pressure vessel applications have a carbon content of less than 0.4%
Steels with carbon content over 0.4% are very brittle and hard to weld
Cast iron use is limited to completed components and configurations because it is
- Very brittle
- Cannot be rolled, drawn, or welded

ASME Code, Section VIII, Division 1 imposes limitations on the pressure and temperature ranges and their repair methods

ASME Code, Section VIII, Division 1 Categories

Carbon Steels

Widely used in pressure vessels
Silicon and manganese are the main alloying elements
Limited to application temperatures below 1000°F

Low-Alloy Steels

Essentially chromium (up to 10%), molybdenum, and nickel alloy steels
These elements enhance the steel for high-temperature applications and in hydrogen service

High-Alloy Steels

Commonly referred to as stainless steels

Consist mainly of chromium (over 10%), nickel, and molybdenum alloys

The three basic types of stainless steels used in process equipment follow

Martensitic Stainless Steels

Group includes type 410, which has low chromium content, slightly above 12%

Behave like steel; are magnetic, heat treatable, and difficult to fabricate

Ferritic Stainless Steels

Group includes type 405 and 430

Magnetic but not heat treatable

Austenitic Stainless Steels

Group includes all 200 and 300 series and contains chromium-nickel and chromium-nickel-manganese steels

Nonmagnetic and not heat treatable

Steel alloys can be produced with a wide variety of alloying elements. Some of the common elements and their effect on steel products are shown on the following table:

Effect of Alloying Elements

ELEMENT	ADVANTAGES
Aluminum	Restricts grain growth
Chromium	Increases resistance to corrosion/oxidation Increases hardenability Adds strength at high temperatures
Manganese	Counteracts sulfur brittleness Increases hardenability

ELEMENT	ADVANTAGES
Molybdenum	Raises grain-coarsening temperature Counteracts tendency toward temper brittleness Enhances corrosion resistance
Nickel	Strengthens annealed steels Toughens steels
Silicon	Improves oxidation resistance Increases hardenability Strengthens steel
Titanium	Prevents formation of austenite in high-chromium steels Prevents localized depletion of chromium in stainless steel during long heating
Vanadium	Increases hardenability Resists tempering

▶ HEAT TREATMENT OF STEELS

General Considerations

The lattice structure of steel varies from one form to another as the temperature changes

Refer to Figure 2.7

Between room temperature and 1333°F, the steel consist of ferrite and pearlite

Ferrite is a solid solution of a small amount of carbon dissolved in iron

Pearlite, shown in Figure 2.8, is a mixture of ferrite and iron carbide

The carbide is very hard and brittle

Between lines A (lower critical temperature) and A_3 (upper critical temperature), the carbide dissolves more readily into the lattice, which is now a mixture of ferrite and austenite

Austenite is a solid solution of carbon and iron that is denser than ferrite

Above line A_3, the lattice has uniform properties, with the austenite the main component; the minimum temperature

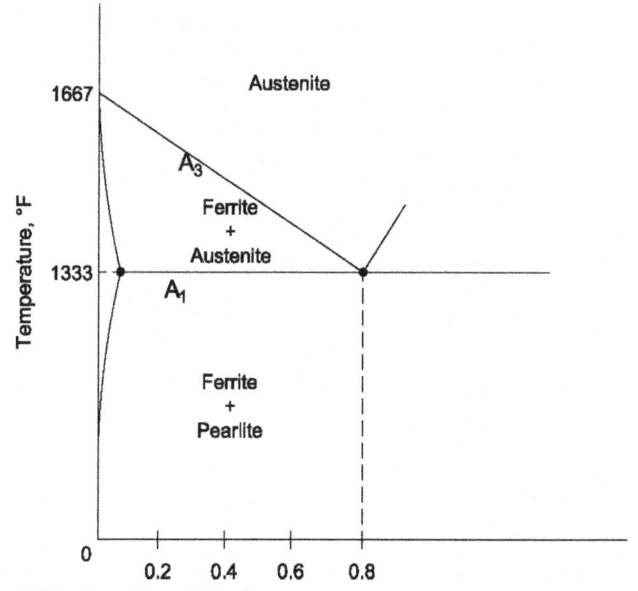

Figure 2.7 Iron-iron carbide equilibrium diagram.

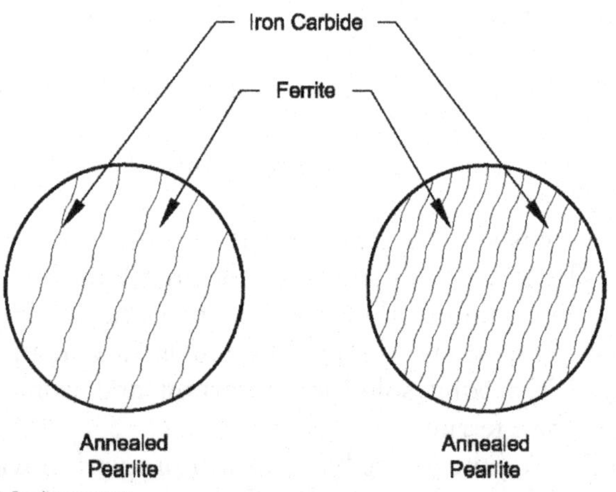

Figure 2.8 Pearlite structure.

for this austenite range is a function of the carbon content of the steel

Various Heat Treatments of Carbon Steel

Normalizing

Consists of heating the steel to about 100°F above the upper critical line A_3 and then cooling in still air

Purpose is to homogenize the steel structure and produce a harder steel than in the annealed condition

Annealing

Consists of heating the steel to about 50°F above the upper critical line A_3 and then furnace-cooling slowly

Purpose is to refine the grain and induce softness

Post-Weld Heat Treating (PWHT)

Consists of heating to a temperature below the lower critical temperature line A_1

Purpose is to reduce the fabrication and welding stress and soften the weld heat-affected zones

Quenching

The rate of cooling of steel after heat treating is very important for establishing the hardness of the steel

Some steels, such as SA-517, obtain most of their high strength by quenching

Rate of cooling depends on many factors:

- Quenching medium
- Temperature
- Size and mass of the part

Tempering

Quenched steels are very brittle

To increase toughness, they are heat treated below A_1 and then cooled to produce the desired property of strength and good toughness

▶ BRITTLE FACTORS

General Considerations

Pressure vessel components constructed of ferrous alloys occasionally fail during the following situations:
- Hydrotest
- Initial start-up
- Normal operating temperatures at a pressure well below the designed value

Such failures occur at low temperatures and could be minimized by incorporating brittle-fracture considerations at the design stage

Degree of sophistication required varies from simple traditional methods to the most complicated mathematical analysis

Application depends on the following factors:
- The amount of information available
- The required reliability of a given component

Both extremes are useful to the designer

Charpy V-Notch Test (C_v)

Simplest and most popular method of qualitatively determining the fracture toughness of low-carbon steels

Test procedure is detailed in ASTM A-370

Procedure consists of the following:
- Impact-testing a notched specimen taken from a specific location of a product form (refer to Figure 2.9A)
- Specimen is struck with a falling weight (Figure 2.9B)

Energy required to fracture it at various temperatures is recorded

Figure 2.10 shows two typical plots of the temperature versus absorbed energy

Significant factors to consider when evaluating material toughness are

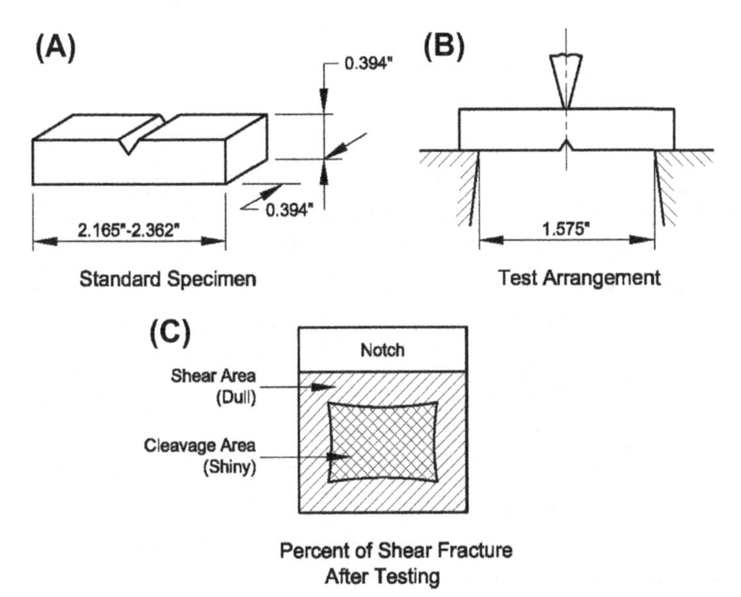

Figure 2.9 Charpy V-notch specimen.

Magnitude of the measured energy
Shape of the energy curve
Appearance of the cross section of tested specimens

The energy level at a given temperature varies with different steels

Refer to ASTM A-593 for guidelines

Energy level of 15 ft-lb at room temperature is adequate for A-283 steel

Same level is exceeding low for A-387 steels

Imperative that the designer specifies the appropriate energy requirements for various steels at different temperatures

Slope of the energy curve in Figure 2.10 gives the rate of change of steel toughness with temperature

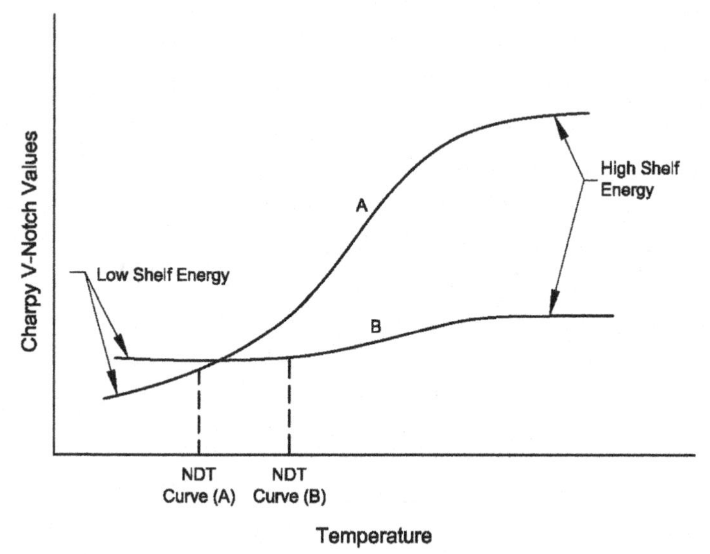

Figure 2.10 C_V energy transition curves.

At the bottom of the curve, the steel is very brittle, as indicated by the cleavage of the tested specimen; failure is normally abrupt

At the upper shelf, material fails in shear and the cross section shows a dull area; failure occurs after excessive yielding

Low-strength steels show a sharp increase in toughness as the temperature increases, as shown in curve A

Higher-strength steels show a slight increase, as shown by curve B

This slight increase in toughness makes the C_V test impractical to use in high-strength steels

The percentage of dull and bright areas in the cross section of tested specimens at a given temperature meets the following characteristics:

Is the measure of the ductility at failure (Figure 2.9C)

- Is helpful in comparing the ductility of two steels at a given temperature
- Is helpful in determining the magnitude of the test temperature with respect to the Nil ductility temperature

The Nil ductility transition (NDT) temperature is of importance when considering low-strength steels
Below this temperature, the fracture appearance of steel changes from part shear to complete cleavage
Thus, vessels of low-strength steel must not be operated below this temperature without a detailed fracture evaluation
The C_v tests give a good qualitative indication of fracture trends
They do not give any correlation between energy and stress levels
Such information is needed where a stress analysis is required; for this reason, other methods were devised, such as the drop-weight test (DWT)

Drop-Weight Test (DWT)

Test procedure is given in ASTM E-208
Procedure consists of the following:

- Welding a brittle bead on a test specimen
- Bead is then notched and the specimen impact tested at various temperatures
- The NDT temperature is obtained when the specimen does not break upon impact

In testing the specimens, deflection can be limited so that the stress at failure does not exceed the yield value; thus, a direct correlation is established between the NDT temperature and yield stress
Such information is used in constructing the fracture analysis diagram (FAD)

Fracture-Analysis Diagram (FAD)

One of the earliest applications of brittle-fracture rules to fail-safe designs

Results obtained from the curve are very conservative but require the minimum of engineering analysis

Figure 2.11 represents a simplified version of the diagram for low-strength steels

Indicates the types of tests required to construct the diagram

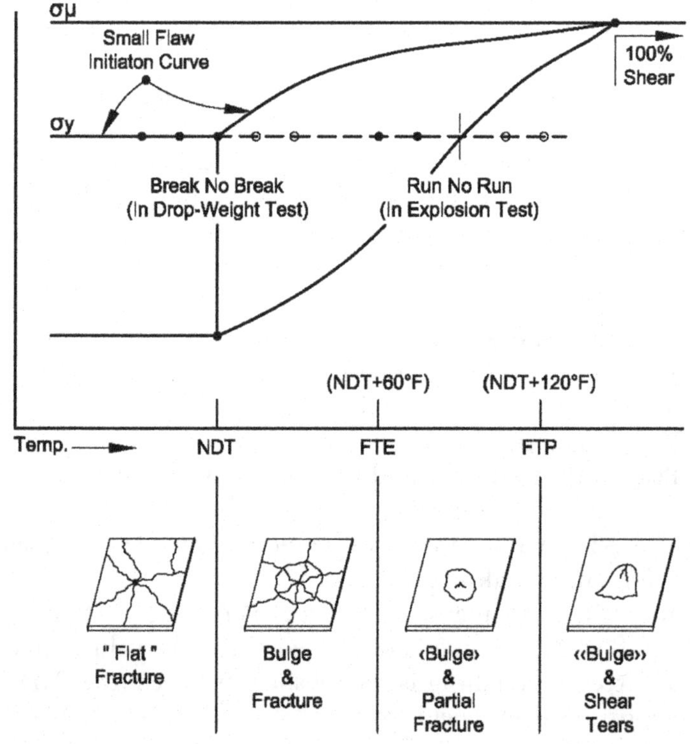

Figure 2.11 Fracture-analysis diagram.

Point A is obtained from the DWT, and it establishes the location of the NDT temperature with respect to yield stress
The crack-arrest temperature (CAT) curve is obtained by running explosive tests on sample plates at various temperatures and observing the crack pattern
Fracture-tear elastic (FTE) temperature point
Obtained when the crack pattern changes from bulge and fracture to bulge and partial fracture
Locates the yield stress with respect to temperature
Fracture-tear plastic (FTP) temperature point
Obtained when the crack pattern changes from bulge and partial fracture to bulge and shear stress
Locates the ultimate stress with respect to temperature
Below point B, fracture does not propagate regardless of the temperature as long as the stress is below 5 to 8 ksi
Between points A and B, other stress lines are drawn to correlate various stress levels
These lines are obtained from the Robertson test
Consists of impact-testing a specimen that is stressed to a certain level and heated from one side to create a temperature, as shown in Figure 2.12
Figure 2.13 shows the complete fracture-analysis diagram
The range of flow sizes at various stress levels has been obtained from experiments as well as experience
Experiments consist of the following:
- Using large spheres of tough material and replacing portions of them with a notched brittle material
- Spheres are then pressurized to a given stress level at the NDT temperature of the brittle material
- The size of the notch is varied with different stress levels to obtain the range indicated in the figure

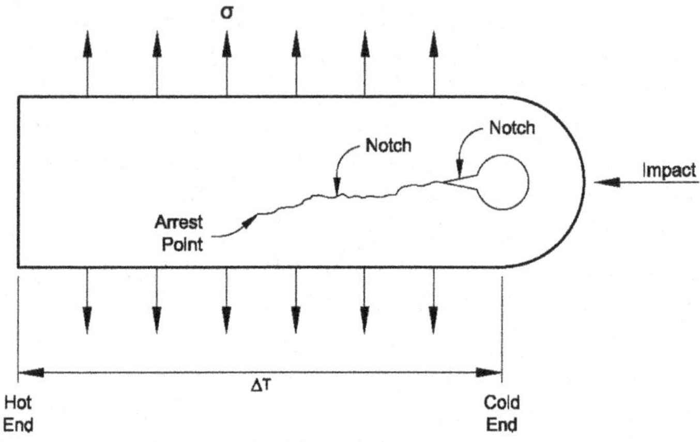

Figure 2.12 Diagram of specimen used in Robertson Crack-arrest test.

The designer must establish the minimum operating temperature of the equipment

Numerous guides are available for determining the minimum temperature at various locations throughout the world

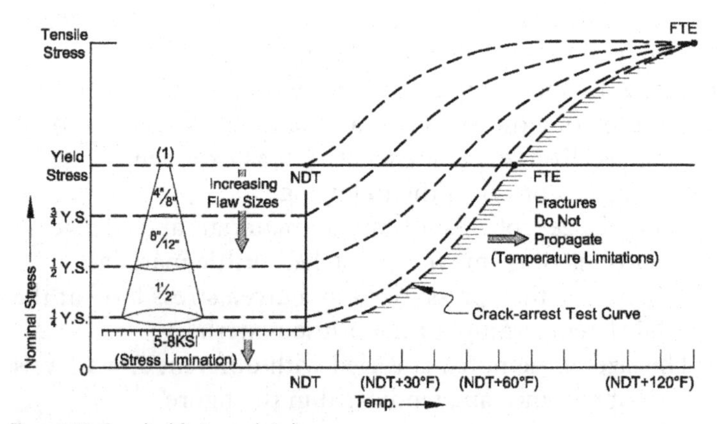

Figure 2.13 Generalized fracture-analysis diagram.

The following limitations must be considered when using Figure 2.13:

It applies only to low-carbon steels

It is valid only for thicknesses of less than 2 inches

Larger thicknesses require special evaluation

It has been proposed that the FTE temperature for thicknesses over 6 inches should be taken as NDT + 120°F rather than NDT + 60°F

The FTP temperature should be NDT + 210°F instead of NDT + 120°F

For thick sections, Figure 2.13 is on the nonconservative side, and the safe operating temperature should be greater than those indicated by the figure

Example 2

Given

A low-carbon steel material with NDT temperature of 15°F is used in a pressure vessel.

Determine

What is the minimum safe operating temperature for such material?

Solution

Because no stress level is given, the minimum stress is assumed at yield.

1. Enter Figure 2.13 at yield stress; the CAT curve is intersected at the FTE point.
2. Moving vertically, a temperature of NDT + 60°F is obtained.
3. Thus, the minimum safe operating temperature is 75°F.
4. If stress concentrations are assumed in the vessel and the stress level is beyond yield at some areas, then a conservative design is at the FTP point.

In this case, the safe operating temperature is NDT + 120°F or 135°F.

Example 3

Given

A low-carbon steel vessel with an NDT temperature of −20°F is to have a start-up temperature of 0°F and a stress level of one-half yield.

Determine

Is the start-up temperature safe?

Solution

1. From the CAT curve in Figure 2.13, the minimum safe temperature is at NPT + 30°F or 10°F for a stress of one-half yield.
2. Thus, the start-up temperature is on the unsafe side because it is less than 10°F; if the start-up temperature is critical, stress will have to be decreased or better impact material selected.

Theory of Brittle Fracture

Assumes that the stress in the vicinity of a crack (Figure 2.14) because of a load applied perpendicular to the direction of crack is given by the following expressions:

$$\sigma_x = \frac{K_I}{\sqrt{2\pi r}}\left(\cos\frac{\phi}{2}\right)\left(1 - \sin\frac{\phi}{2}\sin\frac{3\phi}{2}\right)$$

$$\sigma_y = \frac{K_I}{\sqrt{2\pi r}}\left(\cos\frac{\phi}{2}\right)\left(1 + \sin\frac{\phi}{2}\sin\frac{3\phi}{2}\right)$$

$$T_{xy} = \frac{K_I}{\sqrt{2\pi r}}\left(\sin\frac{\phi}{2}\cos\frac{\phi}{2}\cos\frac{3\phi}{2}\right)$$

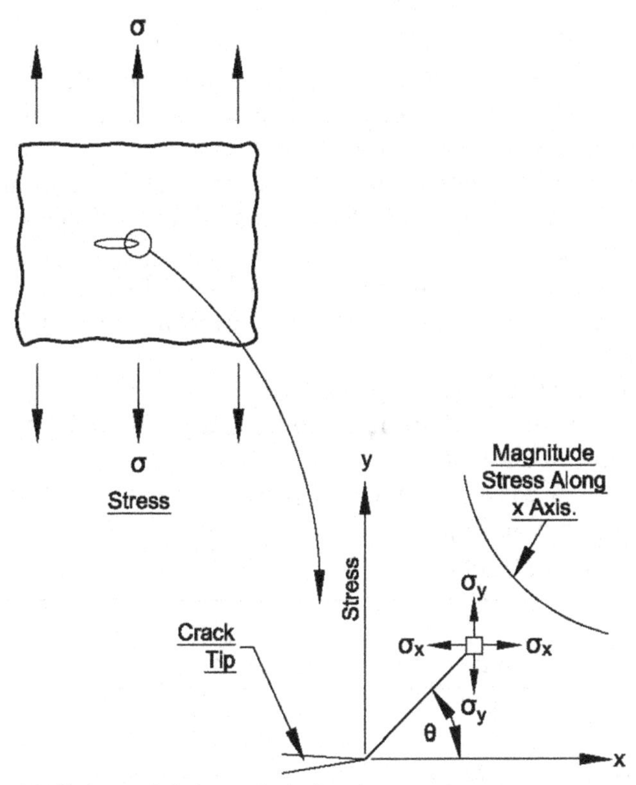

Figure 2.14 Elastic stress distribution near the tip of a crack.

where

σ_x, σ_y, T_{xy} = Stress components at a point, ksi
r, ϕ = Polar coordinates from tip of crack
K_I = Fracture toughness factor (ksi $\sqrt{\text{in}}$)

The fracture toughness factor, K is a function of the applied load as well as the configuration of the body and crack. Thus, K, can be expressed as

$$K_I = \sigma F \tag{1}$$

where

F = Crack shape factor

Unstable crack propagation occurs when the value of K_i reaches a critical value K_{IC}, which is a function of the properties of the material

Temperature variation can have a drastic effect on the value of K_{IC}, as is the case with low-strength carbon steels

K_{IC} values

Experimentally determined; refer to ASTM E-399

Some published values of K_{IC} are given in the following table

Values for the crack shape factor F

Normally obtained from complex analysis based on the theory of elasticity

Only a few cases are suited for practical use; some of them are shown in Figure 2.15

Some Approximate K_{IC} Values

	K_{IC} (KSI $\sqrt{IN}$)			
MATERIAL	−300°F	−200°F	−100°F	0°F
A302, grade B	25	34	48	—
A517, grade F	34	44	77	—
A203, grade A normalized	38	50	—	—
A203, grade A quenched and tempered	42	83	—	—
A55, grade B	35	40	46	78
HY-80	55	—	—	—

In general, materials lose their toughness as the yield strength increases

One measure of toughness is the K_{IC}/σ_y ratio

Ratios > 1.5 indicate tough materials

Case 1: Flow In A Sheet Of Infinite Width.

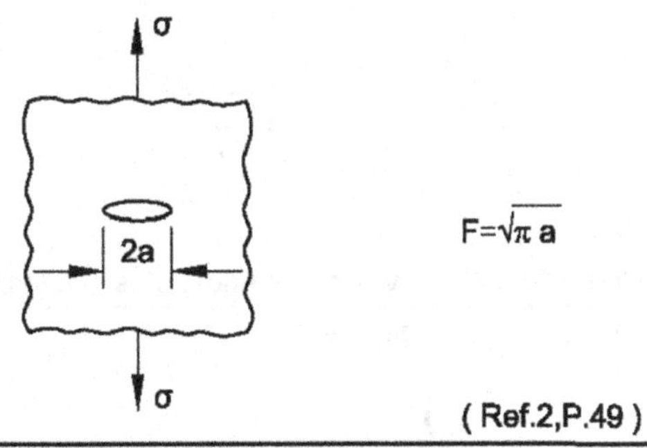

Case 2: Internal Circular Flow In A Sheet Of Finite Width.

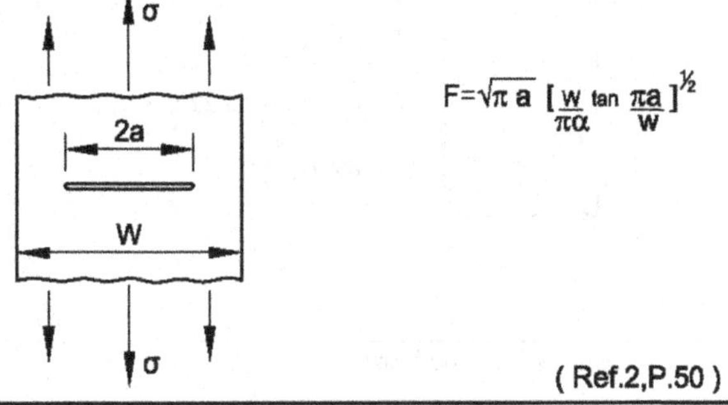

Case 3: Internal Circular Flow In A Thick Plate.

$$F=2\sqrt{\frac{a}{\pi}}$$

Where "A" Is Radius Of Crack (Ref.3,P.39)

Figure 2.15 Shape factors for common configurations.

Case 4: Internal Elliptic Flow In A Thick Plate.

$$F=\frac{\sqrt{\pi a}}{\frac{3\pi}{8}+\frac{a^2}{c^2}}$$

(Ref.3,P.39)

Where "2a" Is The Minor Axis And "2c" Is The Major Axis

Case 5: Single Edge Notch

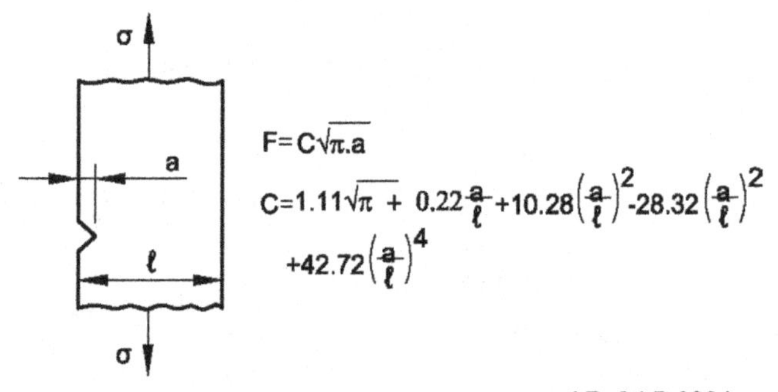

$$F=C\sqrt{\pi.a}$$

$$C=1.11\sqrt{\pi}+0.22\frac{a}{\ell}+10.28\left(\frac{a}{\ell}\right)^2-28.32\left(\frac{a}{\ell}\right)^2+42.72\left(\frac{a}{\ell}\right)^4$$

(Ref.4,P.328)

Case 6: Elliptical Surface Flow.

$$F=\frac{1.12\sqrt{\pi.a}}{\sqrt{\left(\frac{3\pi}{8}+\frac{\pi}{8}\frac{a^2}{c^2}\right)^2-0.212\,\sigma^2/\sigma_y^2}}$$

Where "2c" Is Crack Length, "a" Is Crack Depth, σ Is Actual Material Stress, And σy Is Yield Stress. (Ref.5,P.315)

Figure 2.15 *(Continued).*

Ratios < 7.5 indicate more brittle materials

A study of K_{IC}/σ_y and equation 1 indicates that the defect factor F has to be very small when σ_y is high and K_{IC} is low; in other words, very small defects in high-strength materials can lead to catastrophic failures

Fracture Theory

One of the most accurate methods available for evaluating the maximum tolerable defect size

Main drawback is the difficulty of obtaining K_{IC} for different materials

Economics might dictate a simplified approach like FAD or the ASME criteria with a small permissible defect size rather than a fracture theory approach that might allow a larger tolerable defect

Hydrostatic Testing

Best available method for determining the maximum tolerable defect size

If a thick pressure vessel is hydrotested at a pressure that is 50% greater than the design pressure, the critical K, is given by equation 1 as

$$K_{IC} = \sigma F$$

Assuming an internal defect represented by case 3, the "Shape Factors for Common Configurations" table, the maximum K_{IC}, immediately after hydrotesting is

$$K_{IC} = 1.5 S_M \left(2\sqrt{\frac{A}{\pi}} \right)$$

Maximum defect size x at the design pressure is given by

$$1.5S_M\left(2\sqrt{\frac{\alpha}{\pi}}\right) = S_M\left(2\sqrt{\frac{x}{\pi}}\right)$$

or

$$x = 2.25\alpha$$

Hence, a crack that is discovered after hydrotesting can grow 2.25 times its original size before causing failure

This fact illustrates the importance of hydrotesting; a hydrostatic temperature should be used that is the same as the lowest operating temperature of the vessel

Factors Influencing Brittle Fracture

Many factors affect the brittle-fracture behavior of metals and should be considered in fabricating pressure vessels. Some of the more common are the following.

Torch Cutting or Beveling of Plate Edges

May lead to hard and brittle areas

In cases where this condition is undesirable, the plate should be heated to minimize the effect

Grinding the edges eliminates the hard surfaces

Arc Strikes

Can create failure by brittle fracture, especially if the strike is made over a repaired area

Desirable to grind and repair all arc strikes before hydrotesting, especially at low temperatures

Cold Forming of Thick Plates

May lead to fracture in areas with stress raisers or plate scratches

All stress raisers should be ground off to minimize their effect

Hot forming substantially improves the situation because it increases the NDT temperature and thus prevents brittle fracture

Example 4

Given

A titanium pipe (ASTM B265; Grade 5) with a 2.375-in. Outside Diameter (OD) and a 0.154-in. wall thickness has an actual stress of 30 ksi, a yield stress of 120 ksi, and $K_{IC} =$ 40 ksi $\sqrt{in.}$ at a given temperature. The pipe contains a flaw of depth of 0.05 in. and a length of 0.25 in, which is similar to case 6, represented in the "Shape Factors for Common Configurations" table.

Determine

What is the maximum internal pressure the pipe can hold?

Solution

1. From conventional strength of material analysis, the pressure required to yield the pipe is as follows:

$$P = \frac{\sigma(R_0^2 - R_i^2)}{R_0^2 + R_i^2} = \frac{120(1.188^2 - 1.034^2)}{1.188^2 + 1.034^2} = 16.5 \text{ ksi}$$

2. Using the fracture-toughness approach, the maximum stress is a = K_{IC}/F.
3. From case 6 in the table,

$$F = \frac{1.12\sqrt{\pi(0.05)}}{\sqrt{(1.178 + 0.063)^2 - 0.212(30/120)^2}} = 0.359$$

4. Hence,

$$R = 40/0.359 = 111.4 \text{ ksi}$$

and

$$Max\ P = \frac{111.41(1.188^2 - 1.034^2)}{1.188^2 + 1.034} = 15.4\ ksi$$

Therefore, fracture-toughness criteria controls the design.

ASME Code, Section VIII, Division 1 Criteria

Uses a simplified approach for preventing brittle fracture

Uses exemption curves (Figure 2.16) to determine the following:

Minimum acceptable temperature for a given material

Thickness where impact testing is not required

Curves based on experience as well as test data

Notes to the figure list only a small number of specifications that are assigned to the various exemption curves

Figure requires impact testing

All welded thicknesses > 4 in.

All nonwelded thicknesses > 6 in.

All temperatures < −50°F

The 0.39- inch cutoff limit on the left-hand side corresponds to 1 cm, which is the size of a Charpy V-notch specification

In using the curves, the designer must specify the minimum design metal temperature at which the vessel is to operate, as the code does not provide minimum temperature charts

When a material is required to be impact-tested in accordance with the ASME exemption curves (Figure 2.16), the specified minimum energy level can be obtained from Figure 2.17

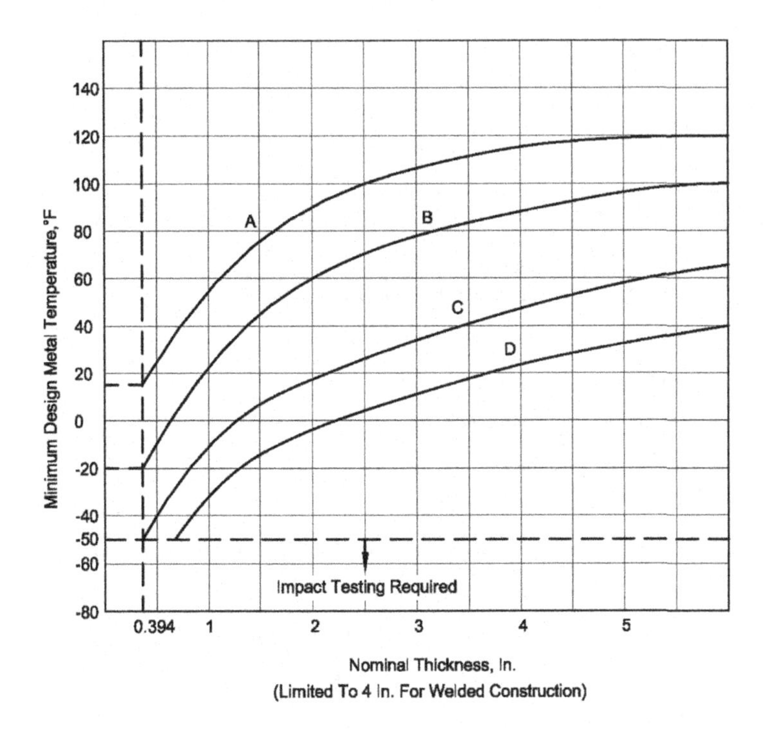

Figure 2.16 Impact exemption curves. Assignment of material to curves.

Accounts for the interaction among the following factors:
- Thickness
- Yield strength
- Toughness

The energy level is obtained from experience with various materials and thickness

Experience indicates that vessels do not fail in brittle fracture when constructed of low-carbon steel or when stress levels are below 600 psi

GENERAL NOTES ON ASSIGNMENT OF MATERIALS TO CURVES:

a. Curve A - all carbon and all low alloy steel plates, structural shapes, and bars not listed in Curves B, C, and D below.

b. Curve B

   1. SA-285 Grades A and B

      SA-414 Grade A

      SA-442 Grade 55 > 1 in. if not to fine grain practice and normalized

      SA-442 Grade 60 if not to fine grain practice and normalized

      SA-515 Grades 55 and 60

      SA-516 Grades 65 and 70 if not normalized

      SA-612 if not normalized

      SA-662 Grade B if not normalized;

   2. all materials of Curve A if produced to fine grain practice and normalized which are not listed for Curves C and D below;

   3. except for bolting [see (e) below], plates, structural shapes, and bars, all other product forms (such as pipe, fittings, forgings, castings, and tubing) not listed for Curves C and D below;

   4. parts permitted under UG-11 shall be included in Curve B even when fabricated from plate that otherwise would be assigned to a different curve.

c. Curve C

   1. SA-182 Grades 21 and 22 if normalized and tempered

      SA-302 Grades C and D

      SA-336 Grades F21 and F22 if normalized and tempered

Figure 2.16 *(Continued).*

SA-387 Grades 21 and 22 if normalized and tempered

SA-442 Grade 55 ≤ 1 in. if not to fine grain practice and normalized

SA-516 Grades 55 and 60 if not normalized

SA-533 Grades B and C

SA-662 Grade A;

2. all material of Curve B if produced to fine grain practice and normalized and not listed for Curve D below.

d. Curve D

SA-203

SA-442 if to fine grain practice and normalized

SA-508 Class 1

SA-516 if normalized

SA-524 Classes 1 and 2

SA-537 Classes 1 and 2

SA-612 if normalized

SA-662 if normalized

e. For bolting the following impact test exemption temperature shall apply.

Impact Test

Spec. No.	Grade	Exemption Temperature, °F
SA-193	B5	-20
SA-193	B7	-40
SA-193	B7M	-50
SA-193	B16	-20

Figure 2.16 *(Continued)*.

SA-307	B	-20
SA-320	L7, L7A, L7M,	Impact tested
L43		
SA-325	1, 2	-20
SA-354	BC	0
SA-354	BD	+20
SA-449	...	-20
SA-540	B23/24	+10

f. When no class or grade is shown, all classes or grades are included.

g. The following shall apply to all material assignment notes.

1. Cooling rates faster than those obtained by cooling in air, followed by tempering, as permitted by the material specification, are considered to be equivalent to normalizing or normalizing and tempering heat treatments.
2. Fine grain practice is defined as the procedures necessary to obtain a fine austenitic grain size as described in SA-20.

NOTE:

(1) Tabular values for this Figure are provided in Table UCS-66.

Figure 2.16 *(Continued)*.

▶ HYDROGEN EMBRITTLEMENT

There are two ways in which hydrogen can cause embrittlement of steels.

Hydrogen Decarburization

Hydrogen penetrates the steel and combines with the carbides in the structure (see Figure 2.7) to form methane gas

Gas accumulates in the space of the original carbide and builds up pressure that leads to cracking

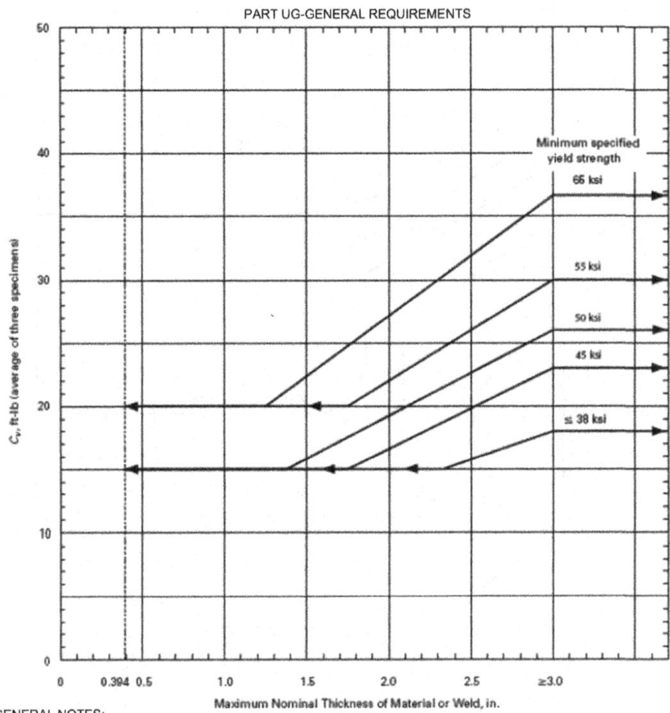

GENERAL NOTES:

(a) Interpolation between yield strengths shown is permitted.

(b) The minimum impact energy for one specimen shall not be less 2/3 of the average energy required for three specimens. The average impact energy value of the three specimens may be rounded to the nearest ft-lb.

(c) Material produced and impact tested in accordance with SA-320, SA-333, SA-334, SA-350, SA-352, SA-420, impact tested SA/AS 1548 (£ impact designations), SA-427, SA-540 (except for materials produced under Table 2, Note 4 in SA-540), and SA-765 do not have to satisfy these energy values. They are acceptable for use at minimum design metal temperature not colder than the less temperature when the energy values required by the applicable specification are satisfied.

(d) For materials having a specified minimum tensile strength of 95 ksi or more, see UG-84(c)(4)(b).

Figure 2.17 Charpy V-notch test requirements.

Process normally accelerates with an increase in temperature and in operating pressure

One method of minimizing hydrogen attack is by using Cr-Mo steels

Here the carbides are in solution with the Cr or Mo and do not readily combine with the hydrogen

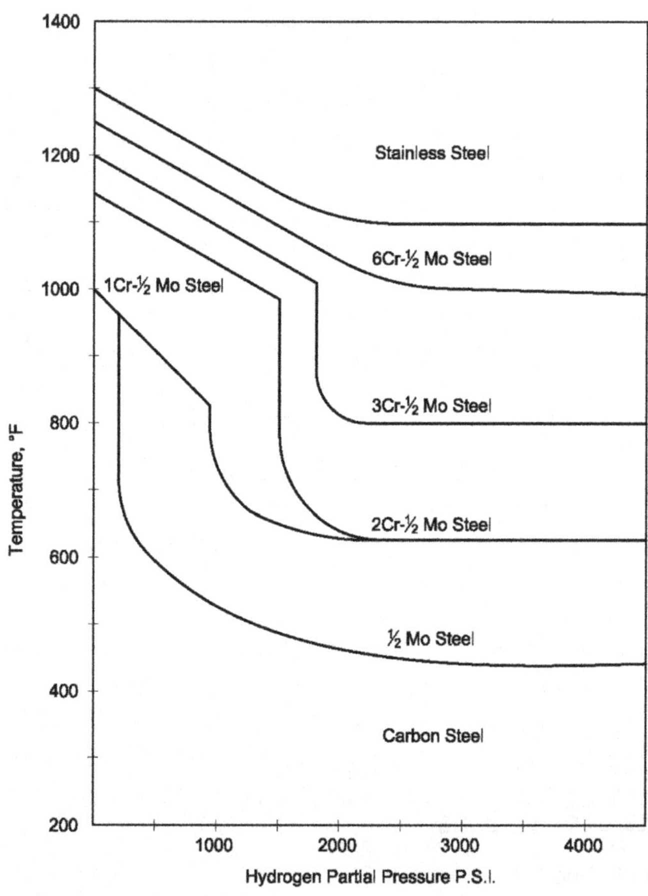

Figure 2.18 The Nelson chart.

The type of steel to be used in a given combination of temperature and pressure is normally determined by the Nelson chart (Figure 2.18)

Hydrogen Attack

Researchers have observed that hydrogen attacks certain regions of a pressure vessel at temperatures below 200°F when they have high hardness zones in the range of 200 Brinell and higher

The exact mechanism is not known, but it is believed that the hydrogen is attached to hard regions with higher-stressed zones

Accordingly, many users require soft heat-affected zones with a Brinell hardness below 200 to avoid hydrogen attack at low temperatures

Materials Selection for Pressure Vessels

3

OVERVIEW

This chapter discusses the following topics:

Selection of materials for service conditions
Brittle fracture prevention
General characteristics of commonly used pressure vessel materials

SELECTION OF MATERIALS FOR SERVICE CONDITIONS

General Considerations

Presentation of the general principles in the selection of materials to prevent deterioration in the service environment

Discusses design factors, typical selections, and characteristics of commonly used materials

Design Factors

Operating temperature and pressure
Service environment
Cost
Design life
Reliability and safety

Operating Temperature and Pressure

Limit the choice of materials
Significantly influence the corrosion rates
Effects of operating temperature:
- Limit materials by adversely affecting strength, metallurgy, and resistance to corrosion

Example, carbon steel is limited to a maximum design operating temperature of 800°F
Above 800°F, the strength of carbon steel decreases significantly and the steel may embrittle because of graphitization
Corrosion rates frequently increase with temperature
Example, in sour services, bare carbon steel is limited to 550°F because corrosion accelerates at higher temperature
Effects of operating pressure:
Influences the stability of a material in the service environment
For example, hydrogen attack of steels in high-pressure, high-temperature H_2S service

Service Environment

Used here means the following:
What the vessel will contain
Its temperature and pressure
Any contaminants
Physical state
Flow rate
Material selection should be based on the following factors:
Corrosion rates
Other potential deterioration mechanisms (stress corrosion cracking and hydrogen damage)
Information about corrosion rates obtained from past experience; a review of the inspection records of vessels in similar service will indicate the following:
Whether material selection was correct
What corrosion rates may be expected
Similarity of new and old service environments
Company "Corrosion Prevention" manuals
Laboratory tests

Published data

Consult a materials specialist for specific recommendations

Cost

Objective is to select the most economical material that will reliably satisfy the design life of the vessel

Normally achieved by doing the following:

Selecting carbon or low-alloy steels in preference to stainless and highly alloyed materials

Specifying conservative corrosion allowances

When stainless steel or a more highly alloyed material is required, it is preferable to use carbon or low-alloy steel clad with a thick layer of the high-alloy material

Clad plate is less expensive, unless the vessel thickness is less than 3/8 to 1/2 inch

Clad plate is less likely to develop though wall stress corrosion cracks than solid alloy

Some commonly used cladding materials (Types 405 and 410 stainless steels) are not practical to fabricate for solid wall construction because of the difficulty in making reliable welds

For some services, up to about 200°F, nonmetallic think-film coatings can be applied to reduce corrosion rates and the need for alloy material

Design Life

Usually 20 years

Exceptions:

Small vessels less than 400 ft^3; if the vessel is easily accessible, a design life of 10 years may be appropriate

Large heavy-walled vessels, thicker than 2 inches; a 30-year design life is recommended

Corrosion allowances

Specified to achieve the design life

Based on expected corrosion rate
Recommendations for pressure vessels are summarized in the following table
If the corrosion allowance required to achieve the design life is greater than ¼ inch, then a more corrosion-resistant alloy or clad vessel is generally economical

Typical Corrosion Allowances for Pressure Vessels

TYPE OF VESSEL	RECOMMENDED MINIMUM (IN INCHES)
Large heavy-walled vessels made of carbon and low-alloy steels	3/16[1]
Carbon and low-alloy steel vessels	1/8[2]
Stainless steel or high-alloy vessels	1/32[3]

[1]If clad, a 0.10-inch minimum cladding thickness is specified to minimize fabrication problems. In this case, no additional corrosion allowance is necessary for the carbon or low-alloy steel.
[2]Usually ⅛ inch is usually used, unless available corrosion data clearly show a corrosion rate less than 3 ml per year. Water legs on drums normally should have a 3/16-inch minimum corrosion allowance.
[3]Applies to solid equipment only. For cladding, a 0.10-inch minimum cladding is specified. Refer to note 1.

Reliability and Safety

Likelihood and consequence of failure must be considered.

Likelihood of Failure

Past history in same or similar services
Whether on-stream inspection can predict failures
Shutdown frequency

Consequence of Failure

Personnel safety (acids, caustic, H_2S, HF, etc.)
Fire hazards (Liquified Petroleum Gas (LPG), high pressure H_2, proximity to furnace)
Lost production (facility/plant profitability)

Ease of repair or replacement
Geographic factors (availability of expert craftsmen and replacement material)
Will leakage cause catalyst poisoning or affect facility/plant performance?
Will facility/plant be shut down, or can equipment be bypassed?
Will facility/plant cause force-related facility/plant shutdowns?
Will leakage cause environmental problems such as pollution of navigable waters?
Considerations of these factors may lead to conclusions that differ from the guidelines contained herein and thus warrant different materials or corrosion allowances
Each individual case warrants consideration of these factors, and judgment is then necessary to choose economical materials

Typical Selections

The following table illustrates pressure vessel materials typically selected for common environments
Not suitable for final materials selection
May save time in initial investigation

Common Pressure Vessel Materials

SERVICE	TYPICAL MATERIALS	COMMENTS	NOTES
Produced fluids containing water	Carbon steel with coating	Corrosivity of produced fluids varies widely	1
Sweet hydrocarbons, less 1 ppm H_2S	Carbon steel	May corrode even with trace H_2S above 550°F	1

(Continued)

Common Pressure Vessel Materials

SERVICE	TYPICAL MATERIALS	COMMENTS	NOTES
Sour hydrocarbons, with more than 1 ppm H_2S below 550°F	Carbon steel	Limited to 550°F maximum	1
Sour hydrocarbons, with more than 1 ppm H_2S above 550°F	Carbon steel clad with 12% Cr stainless steel		2
Sweet hydrogen, such as in catalytic reformers, hydrogen and ammonia plants	Carbon steel, 1¼ Cr-½ Mo and 2¼ Cr Mo steels	Choice depends on temperature and hydrogen partial pressure; see API Recommended Practice 941	1, 3
Sour hydrogen; may also contain hydrocarbon; examples include hydroprocessing unit and hydrotreater process streams	Carbon steel, 1¼ Cr-½ Mo, and 2¼ Cr-1 Mo Often clad with Type 321 or 347 stainless steel	Choice depends on temperature and on hydrogen and H_2S partial pressure; see API Recommended Practice 941	4
Steam	Carbon steel	CO_2 corrosion may demand stainless cladding in a condensation service	1
Amines (MEA, DEA)	Carbon steel	Stress-relieve new pressure vessels to prevent stress corrosion cracking; stainless steel cladding is frequently used in selected areas, such as in regenerators, to minimize corrosion; consult a materials specialist	1
Caustic (<200°F)	Carbon steel	To prevent stress corrosion cracking,	1

Common Pressure Vessel Materials

SERVICE	TYPICAL MATERIALS	COMMENTS	NOTES
		stress relief is required as follows: (1) for caustic with a concentration less than 30 weight percent, stress-relieve in service above 140°F; (2) for caustic with a concentration of 30% or greater, stress-relieve in service above 110°F	
Sulfuric Acid (≥85% concentration)	Carbon steel	Velocity above about 3 fps and temperature above 120°F will result in severe corrosion of carbon steel; vessels handling sulfuric acid and LPG mixes, such as in alkylation plants, usually are made of carbon steel	1
Sour water	Carbon steel	Carbon steel may corrode at high concentrations of NH_3 and H_2S	

[1]Carbon steel. Grades commonly used for pressure vessel plates are SA 285 grade C, SA 515 Grade 70, and SA 516 Grade 70. Choice will be determined by minimum design metal temperature and thickness.

[2]Clad carbon steel. Carbon steel clad with 12% Cr steel is covered by specification SA 263. We usually designate a base metal plate (carbon steel per note 1 above) and the cladding as Type 405 or Type 410S.

[3]Low-alloy steels. 1¼ Cr-½ Mo steel is covered by SA 387 Grade 11 (plate) and SA 336 F11 (forgings).

[4]Carbon or low-alloy steel clad with Type 321 or 347 stainless steel. These plates are covered by SA 264 for roll band cladding. Base metal plate is designated per notes 1 or 3 above. If forgings are used for shell components of if shell plates are thick, they will be weld overlay clad rather than roll band clad. Base metal will be designated per notes 1 and 3 above.

Application Criteria for Common Pressure Vessel Materials

Carbon Steel

Readily available and easily fabricated

Economical material selection, with ⅛-to ¼-inch corrosion allowance

Nominal composition of iron with about 1% manganese and up to 0.35% carbon

Higher carbon results in poor weldability

Some limitations are as follows.

Brittle Fracture

Carbon steels may be susceptible to brittle fracture at normal ambient temperatures

Refer to "Selection of Materials for Brittle Fracture Prevention" at the end of this section

Hydrogen Attack

Carbon steel will suffer hydrogen attack at elevated temperature in high-pressure hydrogen

Material selection should be based on the Nelson curves

Refer to API RP 941 for additional information

Graphitization

Formation of graphite, primarily in weld heat-affected zones, from the decomposition of iron carbides

Graphitized steel can fail under small loads or strains

Welded carbon steel is limited to 800°F maximum

Stress Corrosion Cracking (SCC)

Welded or cold-worked carbon steel is susceptible to SCC in caustic, nitrate, carbonate, amine solutions and in anhydrous ammonia

Stress relief is required to prevent failures

Consult a materials specialist for specific applications

Sulfide Stress Cracking (SSC)

High-strength steel and hard welds in steel in aqueous solutions containing H_2S and susceptible to sudden nonductile failures

Controlling maximum strength and hardness is generally sufficient to prevent cracking

Post-weld heat treatment may be beneficial to prevent cracking

Hydrogen-Induced Cracking (HIC)

Some low-strength carbon steels may be susceptible to HIC in wet services containing H_2S

Blistering is one example of this type of cracking

Stress-oriented, hydrogen-induced cracking (SOHIC) is a specialized type of HIC that has in some cases resulted in through-wall cracks in carbon steel pressure vessels

Manufacturers offer steels made with very low sulfur contents and calcium treated for inclusion shape control to resist HIC

Standard tests are available to evaluate the HIC resistance of steel plates

Post-weld heat treatment may also be beneficial to prevent cracking

Most companies do not traditionally specify these steels

Carbon-Molybdenum Steel

Similar to carbon steel but with 0.5% molybdenum added

Molybdenum improves the steel's high-temperature strength and graphitization resistance

Corrosion resistance is the same for carbon steel

Some limitations

Brittle Fracture

Unless made to fine-grain practice and normalized, may have poor toughness (increased susceptibility to brittle fracture)

Hydrogen Attack

Experience indicates they cannot be relied on to resist hydrogen attack

Should not be specified for hydrogen attack resistance but instead 1¼ Cr to 1½ Mo should be specified

Refer to API RP 941 and 571 for additional information

Graphitization

Like carbon steel, carbon-molybdenum steel will graphitize, but it is resistant to a maximum service temperature of 850°F

Stress Corrosion Cracking

Same as carbon steel

Sulfide Stress Cracking

Same as carbon steel

Chrome-Molybdenum Steel

Similar to carbon steel but with chromium and molybdenum added

Typical grades are Cr-½ Mo, 1¼ Cr-½ Mo, and 2¼ Cr-1 Mo

Corrosion resistance is about equal to that of carbon steel

Offers better resistance to hydrogen attack and better high-temperature strength

Do not graphitize

More difficult to fabricate and requires the following:

Control of preheat for welding
Post-weld heat treatment for all welded construction

Some limitations include the following.

Brittle Fracture

Like carbon steels, undergo a ductile-to-brittle transition at low temperatures and become susceptible to brittle fracture
Embrittlement in service above 650°F
The 2¼ Cr-1 Mo steels are particularly susceptible, but 1 Cr-½ Mo and 1¼ Cr-½ Mo may also be susceptible
Screening tests are recommended to minimize embrittlement

Hydrogen Attack

Resistance is dependent on the chromium and molybdenum contents in the steel; resistance improves with increased alloy content
Refer to API RP 941 and 571
Stress corrosion cracking and sulfide stress cracking
Same limitations as for carbon steel

Stainless Steel

Alloys of iron and chromium, typically with at least 12% chromium
300 series steels contain nickel
Common term for Type 304 is 18-8
Translates into 18% chromium – 8% nickel
Other alloying elements such as molybdenum, titanium, and niobium are added for specific purposes
Classified, depending on their microstructure, as follows:
Austenitic
Ferritic

Martensitic
Duplex

Austenitic

Have an austenitic structure similar to the high-temperature structure of carbon steel
Will not harden with heat treatments

Nonmagnetic

Readily weldable and used for both cladding and solid wall construction
Examples are Types 304, 316, 321, and 347

Ferritic

Have a ferritic structure similar to the low-temperature structure of carbon steel
Will not harden with heat treatment
Magnetic and do not contain nickel
Used for cladding (Type 405), as solid construction is limited because of poor weldability
Examples are Types 405 and 430

Martensitic

Can be hardened with heat treatment
Magnetic
Used for cladding, as solid construction is limited due to poor weldability
Most common example is Type 410

Duplex

Have structures of roughly 50% austenitic and 50% ferrite
Nonhardened by heat treatment

Not widely used for pressure vessels but could be used for both cladding and solid wall construction
Have corrosion properties similar to austenitic but are higher in strength
Share some of the limitations of both the ferritic and austenitic

Some limitations include the following.

Austenitic Stainless Steels in Chloride Solutions

Chloride stress corrosion cracking can occur in aqueous solutions containing chloride ions
Cracking is most severe in the following situations:
  Chloride ion concentration is high
  Solution is hot
  pH is neutral or low
Evaporation builds up deposits on the stainless steel
Types 304, 316, 321, 347, and the like are affected
Stainless equipment hydrostatically tested with seawater has failed because of the residual sodium chloride film left behind
Other failures have been traced to the following:
Chlorides leaching out of wet insulation
Not protecting stainless equipment from chlorides during shutdowns
There can be an incubation period of several hours to weeks before cracking occurs in certain environments
Cracking can be reduced by stress relieving the stainless equipment in the 1550°F to 1650°F temperature range
Complete freedom from chloride stress corrosion cracking can only be assured by doing the following:
  Protecting austenitic steels from any chloride ions
  Using more expensive super grades with 30% to 45% nickel

Duplex stainless steels have improved resistance to chloride stress corrosion cracking

Recommendations to prevent chloride stress corrosion cracking include the following:

1. Do not select solid wall austenitic stainless steel construction for hot, aqueous chloride services; if stainless steel is required, use clad construction
2. Stress relieve vessels made of solid austenitic stainless steel where no other economical material is available

Austenitic Stainless Steels in Sulfur-Derived Acids (SDA)

SDA can cause "polythionic acid" stress corrosion cracking

Unlike chloride stress corrosion cracking, the steel must be sensitized with chromium carbide precipitates along the grain boundaries before polythionic acid stress corrosion cracking can occur

Sensitization results from exposure of stainless steel equipment to temperatures in excess of 700°F

If regular carbon grades (Types 304 or 316) are used, they may sensitize during welding

Neither sulfurous nor polythionic acids are normally found in process units during operations

These acids commonly develop during shutdowns by the oxidation of iron sulfide scale in the presence of moisture and oxygen

Also form in flue gas condensate

Freedom from polythionic acid stress corrosion cracking can be assured only by preventing sensitized austenitic stainless steels from coming in contact with sulfur-derived acids

Regular grades (Types 304, 216, etc.) sensitize easily at temperatures above 700°F

The heat of welding is often enough to sensitize the heat-affected zone

Extra low carbon grades of stainless steel (Types 304L, 316L, etc.)

Do not sensitize during welding

Will sensitize with long-term exposure at temperatures above 700°F

Types 321 and 347 are chemically stabilized to minimize sensitization

Polythionic acid cracking is prevented by the following practices:

- Using chemically stabilized or extra low carbon grades of stainless steel
- Avoiding harmful heat treatments

A less effective means of prevention is to use regular grades of stainless steels and required soda ash wash during all shutdowns

Chromium Stainless Steels in 750°F to 900°F

Stainless steels containing more than 13% chromium can suffer embrittlement during exposure to temperatures in the 750°F to 900°F range; occurs in the following types:

- Ferritic (Types 405 and 430)
- Martensitic (Type 410)

Known as "885°F embrittlement"

Some steels are so sensitive that even slow cooling through the temperature range will cause embrittlement

Embrittlement results in an upward shift in the ductile-to-brittle transition temperature

Duplex stainless types are also susceptible

Prevent this problem by not using chromium stainless steels for solid wall construction of pressure vessels

Stainless Steels above 1000°F

At elevated temperature all stainless steels with high chromium contents will develop some *sigma phase,* which causes embrittlement at lower temperatures

Sigma phase

- Is very hard, non-magnetic, and brittle
- Composition varies depending on the alloy from which it formed
- Does not affect the steel's elevated temperature properties but may make it so brittle at lower temperatures that failures will occur during startup or shutdown

Stainless steels containing more than 13% chromium (ferritic and martensitic) are very susceptible to extensive "sigma phase" formation

Austenitic stainless types are not as susceptible because of their high nickel content but can develop damaging amounts of sigma phase when held between 1000°F and 1500°F for long periods of time

Certain highly susceptible austenitic alloys, such as castings and welds, may develop serious embrittlement in a few hours at temperatures of 1200°F to 1300°F

Duplex stainless steels are also very susceptible to sigma embrittlement

Sigma embrittlement is controlled by minimizing ferrite content of stainless steel welds

Duplex stainless steel is limited to 650°F maximum service temperature to avoid embrittlement

Sulfide Stress Cracking

Martensitic stainless steels are especially susceptible

Welds are difficult to soften with heat treatment and are therefore susceptible to cracking

Low carbon grades (Type 410S) are used to limit weld zone hardness

Prevented by controlling weld strength and hardness

Other Alloys

Other alloys are not frequently used for pressure vessel construction. Two classes are occasionally considered.

Nickel Alloys

Examples include Monel, Inconel alloys, Incoloy alloys, Hastelloy alloys (all are very expensive) used for specialized applications and then usually as cladding

Some nickel alloys have good resistance to chloride solutions where stainless steels are poor

Fabricating and weldability are generally good with proper precautions

Titanium Alloys

Used infrequently for pressure vessels

Welding is difficult, requiring very clean conditions

Welding usually done only in a shop "clean room," so field repairs are not practical

Summary of Temperature Limitations

The following table summarizes the applicable properties and temperature limitations of commonly used pressure vessel materials:

Maximum Temperature Limits of Common Pressure Vessel Materials, °F

	CARBON STEEL	C-½ MO	1⅛ CR-1 MO	2¼ CR-1 MO	12 CR (410)	18CR-8NI (304)
Strength* (3000 psi)	990	1075	1135	1150	1100	1275
Oxidation (10 mpy loss)	1025	1025	1050	1100	1350	1600
Graphitization (welded only)	800	850	N/A	N/A	N/A	N/A

(*Continued*)

Maximum Temperature Limits of Common Pressure Vessel Materials, °F

	CARBON STEEL	C-½ MO	1⅛ CR-1 MO	2¼ CR-1 MO	12 CR (410)	18CR-8NI (304)
885 Embrittlement	N/C	N/A	N/A	N/A	775-950	N/A
Sigma embrittlement	N/A	N/A	N/A	N/A	N/A	1100-1700
Hardening on cooling	1330	1330	1375	1425	1450	N/A
Carbide precipitation	N/A	N/A	N/A	N/A	N/A	850-1550
Hydrogen attack H_2pp (750 psi)	500	500	1000	1100	N/A	N/A
Caustic stress corrosion cracking	140	140	140	140	140	140
Chloride stress corrosion cracking	N/A	N/A	N/A	N/A	N/A	140
Sulfide stress cracking	X	X	X	X	X	N/A

X = Susceptible when yield strength exceeds 90 ski or hardness exceeds Rockwell C22.
N/A = Not applicable.
* = 100,000-hour stress rupture strength (typical).

▶ SELECTION OF MATERIALS FOR BRITTLE FRACTURE PREVENTION

General Considerations

Brittle Fracture

Sudden, often catastrophic failure that is inherent to brittle material

Involves little or no deformation (yielding)

Occurs in steel structures such as pressure vessels, tanks, pipes, beams, and so on

Brittleness

Material property that indicates the material is prone to failure without deformation (shatter like glass)

Brittle materials are prone to failure when they are stressed in the vicinity of a notch or stress concentration

Toughness

Opposite of brittleness—that is, a material's ability to resist brittle fracture

Depends on material strength, thickness, and temperature

To resist brittle fracture, higher-strength materials and thick materials require greater toughness than low-strength and thin materials

More Considerations

Brittle fractures can occur in ferritic steels within the normal atmospheric temperature range

Carbon, carbon-½ moly, chrome-moly, and 400 series

Regular 300 series steels

Are not susceptible to brittle fracture until temperatures are below −300°F

After exposure above 1100°F, sigma embrittlement makes weld metals with large amounts of ferritic phase susceptible to fracture well above room temperature

Brittle fractures are infrequent; most occur during hydrotest rather than in operation

Can be catastrophic because of fragmentation of the structure and the fast release of energy

Characterized by a flat fracture surface and occurs at average stress levels below those of general yielding

Brittle fracture cracks grow at very fast speeds (up to 7000 ft/sec), so brittle fracture happens quickly and unexpectedly

Design to Prevent Brittle Fracture

Overview

Susceptibility of structures to brittle fracture depends on the following:
- Preexisting flaw size
- Tensile stress level
- Fracture toughness of the material

Flaw size and stress level are controlled by design, fabrication, and inspection in accordance with the ASME code

Toughness is controlled by material selection in accordance with the ASME code

To prevent brittle fracture, do the following:

Keep flaw size small and stress levels low

Use tough materials

Toughness is a physical property of materials that primarily characterizes their resistance to brittle fracture, depending on temperature, loading rate, and thickness

Sufficiently tough steels are selected by one of the following:
- Using materials selection curves, or impact exemption curves contained in the ASME code
- Using steels that have been Charpy V-notch (C_V) impact tested to code requirements

Using steels selected from the ASME code impact test exemption curves is highly preferred

Steel selection is discussed in more detail later in this section

Background: Transition Temperature Approach to Fracture Control

Ferritic steels (carbon, low-alloy, and 400 series)

All undergo a ductile-to-brittle transition as temperature is lowered

Each have a ductile-to-brittle transition temperature range

Above the range, these steels are tough

At and below the range, they can fracture in a brittle manner
Toughness
Ability of a material to absorb energy by yielding (plastic deformation) prior to failure
Depends on a material's ductility and strength
Indicates the material's ability to resist brittle fracture
One measure of toughness is the area under a tension stress-strain curve taken to failure
The standard method for pressure vessel application is the Charpy V-notch impact test
Measures the energy to fracture a specimen under very high strain rates (sudden impact)
Used to make sure the transition temperature of the steel is below the minimum loading temperature of the vessel (transition temperature approach)
Figure 3.1 illustrates a typical C_V impact test result for carbon steel
The C_V transition temperature is defined as the minimum temperature above which the material requires more than some specified energy to break
Energy required for establishing transition temperature increases with increasing steel strength
Charpy V-notch transition temperature approach
Developed from analysis of ship failures
More than 100 structural failures were analyzed
Fracture initiation was found to be difficult above a transition temperature corresponding to a C_V impact energy of 10 ft-lb
Crack propagation was difficult above a temperature corresponding to 15 to 25 ft-lb
From these findings, a 15 ft-lb C_V requirement at the minimum loading temperatures became a widely used fracture criterion

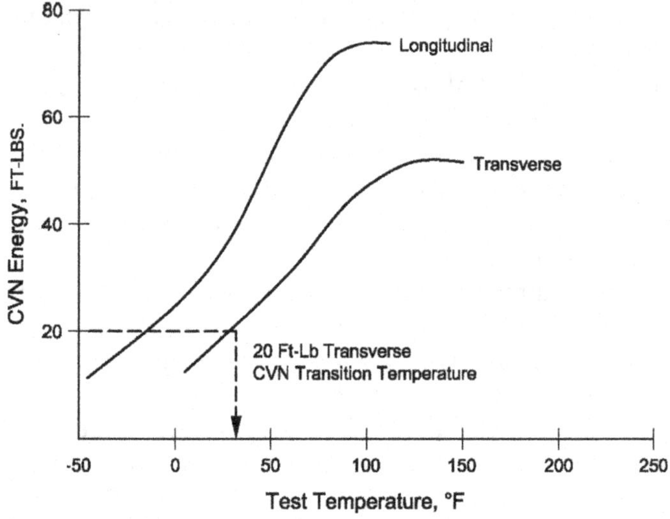

Notes :

1. These Data Illustrate The Variation Of CVN Energy With Temperature And With The Orientation Of Test Specimen Relative To The Direction Of Principal Working.
2. These Data Must Not Be Considered Typical. Wide Variation May Result Even From Specimens From Plates Of The Same Specification And Thickness.

Figure 3.1 Illustration of typical Charpy V-notch impact test data.

ASME Code, Section VIII, Division 1

Requirement for C_v energy for pressure vessel steels are given graphically in Figure UG-84.1

Steels may be exempted from tests if they meet requirements shown in figure UC5-66

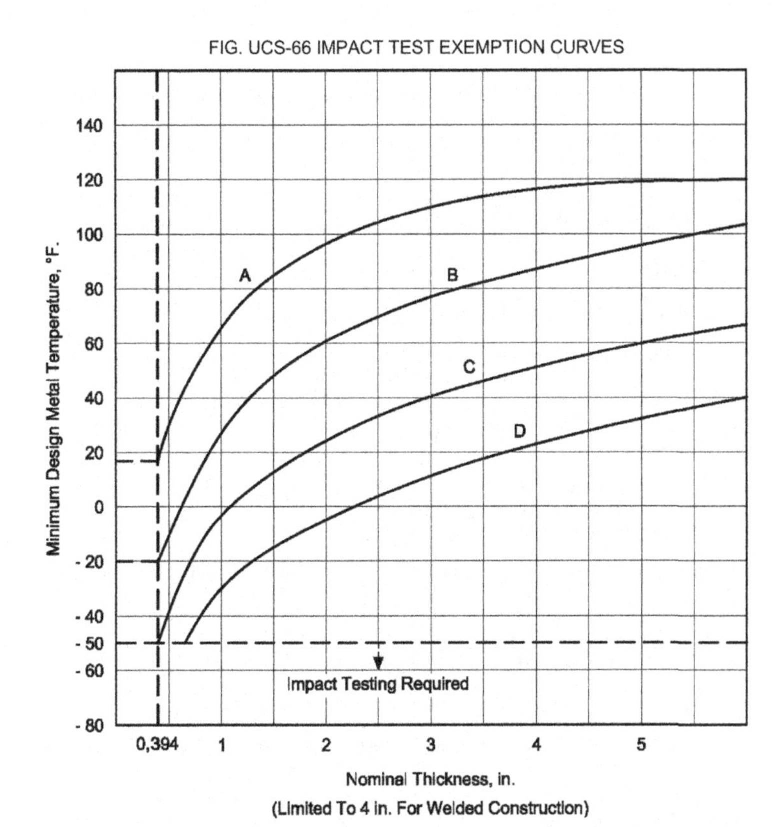

Figure 3.2 Illustration of typical Charpy V-notch impact test Exemption Curves.

GENERAL NOTES ON ASSIGNMENT OF MATERIALS TO CURVES:

(a) Curve A - all carbon and all low alloy steel plates, structural shapes, and bars not listed in Curves B, C, and D below.

(b) Curve B

(1) SA-285 Grades A and B

SA-414 Grade A

SA-442 Grade 55 > 1 in. if not to fine grain practice and normalized

SA-442 Grade 60 if not to fine grain practice and normalized

SA-515 Grades 55 and 60

SA-516 Grades 65 and 70 if not normalized

SA-612 if not normalized

SA-662 Grade B if not normalized;

(2) all materials of Curve A if produced to fine grain practice and normalized which are not listed for Curves C and D below;

(3) except for bolting [see (e) below], plates, structural shapes , and bars, all other product forms (such as pipe, fittings, forgings, castings, and tubing) not listed for Curves C and D below;

(4) parts permitted under UG-11 shall be included in Curve B even when fabricated from plate that otherwise would be assigned to a different curve.

(c) Curve C

(1) SA-182 Grades 21 and 22 if normalized and tempered

SA-336 Grades F21 and F22 if normalized and tempered

SA-387 Grades 21 and 22 if normalized and tempered

SA-442 Grade 55 $\leq$ 1 in. if not to fine grain practice and normalized

SA-516 Grades 55 and 60 if not normalized

SA-533 Grades B and C

SA-662 Grade A;

(2) all material of Curve B if produced to fine grain practice and normalized and not listed for Curve D below.

(d) Curve D

SA-203

SA-442 if to fine grain practice and normalized

SA-508 Class 1

SA-516 if normalized

SA-524 Classes 1 and 2

SA-537 Classes 1 and 2

SA-612 if normalized

SA-662 if normalized

(e) For bolting the following impact test exemption temperature shall apply.

Impact Test

Spec. No.	Grade	Exemption Temperature, °F
SA-193	B5	-20
SA-193	B7	-40

SA-193	B16	-20
SA-307	B	-20
SA 320	L7, L7A, L7M,	Impact tested
SA-325	1, 2	-20
SA-354	BC	0
SA-354	BD	+20
SA-449	...	-20
SA-540	B23/24	+10

(f) When no class or grade is shown, all classes or grades are included.

(g) The following shall apply to all material assignment notes.

(1) Cooling rates faster than those obtained by cooling in air, followed by tempering, as permitted by the material specification, are considered to be equivalent to normalizing or normalizing and tempering heat treatments.

(2) Fine grain practice is defined as the procedures necessary to obtain a fine austenitic grain size as described in SA-20.

NOTE:

(1) Tabular values for this Figure are provided in Table UCS-66.

ASME Code, Section VIII, Division 2

Requirements for C_v energy for pressure vessel steels are given in Table AM-211.1

Requirements are less conservative than Division 1 and do not take into account the need for higher C_v energy with increasing thickness

Steels may be exempted from tests if they meet requirements shown in Paragraph AM-218

Selecting Steels for New Construction of Pressure Vessels

Ensure that the vessel temperature is above a minimum during "loading of the equipment"

Minimum temperature is defined as the "minimum pressurizing temperature"

Loading includes the following:

Operation

Hydrotest

Pressure test

Startup and shutdowns

ASME Code, Section VIII, Division 1

Minimum Pressurizing Temperature (MPT)

Lowest temperatures at which a pressure greater than 40% of the MAWP should be applied to the vessel

Below 40% of the maximum allowable working pressure (MAWP), stresses are considered low enough to essentially eliminate the risk of brittle fracture in the absence of significant other stresses (such as those due to weight and differential thermal expansion)

To establish a minimum design metal temperature, the following should be considered:

Startup temperature

Normal and abnormal operating conditions

Best available local weather data

Lowest 1-day mean temperature shown in Figure 2-2 of API 650

Materials Selection Requirements

One of two methods are used to assure that steels are used above their transition temperature:

Impact test exemption curves

Charpy V-notch impact testing

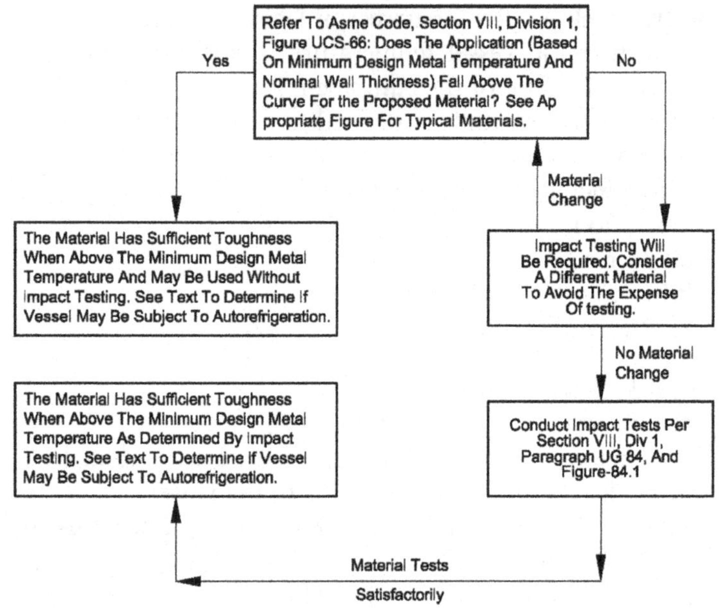

Figure 3.3 Simplified overview of design for brittle fracture under ASME Code, Section VIII, Division 1.

Figure 3.2 is a flowchart applying these methods under the code

Impact-test exemption curves are preferred to C_V impact testing where prior data or service experience is available

C_V tests

Increase material's costs substantially (approximately 2 to 10 cents/pound)

Complicate delivery

Impact Test Exemption Curves: Code Paragraph Figure UCS-66

Gives the application points (combinations of thickness and minimum design metal temperature) where prior data or service experience shows specific steels have sufficient toughness for fracture-safe design (above transition temperature)

The application point is the point corresponding to the thickness and minimum pressurizing or design metal temperature

A steel has adequate toughness if the application point is above the steel's curve (refer to UCS-66)

To use a steel at an application point below that steel's curve, C_v impact testing is required to prove adequate toughness

Most companies prefer to use steel with suitable inherent toughness rather than requiring impact tests, which can cause unnecessary delays and expense

Follow the requirements of Code Paragraph UCS-66 and Figure UCS-66 to do the following:

Select steels for pressure vessels

Establish the need for impact testing

However, the following restrictions on the use of UCS-66 are recommended

1. All grades of SA 285 and SA 515 steels thicker than ¾ inch should be assigned to curve A rather than curve B; this is more conservative than the requirements of Figure UCS-66
2. SA 285 is often semikilled steel and SA 515 is made to coarse-grain practice, so both tend to have poorer impact transition temperatures (higher) than curve B indicates

The following restrictions on the use of UCS-66 are recommended:

The impact test exemption allowed in Code Paragraph UG-20(f) should not be allowed

Paragraph UG-20(f) eliminates the impact test requirements for carbon steel 1 inch or less in thickness for most pressure vessels

It is better to choose steels from figure UCS-66 that do not require testing for that specific application

Note: Code Paragraph UCS-68(c) allows a 30°F reduction in impact testing exemption temperature for P-1 materials (carbon steel) that are given post-weld heat treatment (PWHT) and PWHT is not otherwise required by the code

For example, this reduction would not be allowed for vessels of carbon and low-allow steels that are in "lethal service," as PWHT is mandatory

Charpy V-Notch Impact Testing

When steel is to be used at an application point below its curve on Code Figure UCS-66, C_V impact testing is required to prove adequate toughness

Requirements for C_V testing are summarized below:

1. Each plate, forging, or pipe used at an application point below its impact test exemption curve is tested; usually each plate is tested, whereas forgings and pipe are tested in accordance with specifications such as SA 350 and SA 333
2. Three test specimens taken transverse to the major working direction (during steel making) are tested; it is important that transverse rather than longitudinal specimens be used because transverse properties are generally poorer; seamless pipe is an exception because its properties do not vary much with orientation; ASTM A-370 defines transverse and longitudinal C_V specimens; the code leaves orientation optional; specimen orientation is a company requirement
3. The maximum (warmest) allowable C_V test temperature is the minimum pressurizing or design metal temperature
4. Minimum C_V energy requirements are in accordance with Code Figure UG 84.1. Figure UG 84.1 shows C_V

energy requirement as a function of specified minimum yield strength and thickness

5. When impact testing is required on the parent metals, impact testing of the heat-affected zone (HAZ) and deposited weld metal is required on the welding procedure qualification test plate (WPQT) or production test plate; refer to Code Paragraph UG-84 for definitions of these terms; see also Code Paragraph UCS-85 concerning heat treatment of test specimens; test specimen heat treatment must simulate actual vessel heat treatment

ASME Code, Section VIII, Division 2

Minimum Pressurizing Temperature

Code Paragraph AD-155 defines the minimum permissible vessel metal temperature for ferrous metals other than austenitic—that is, for carbon and low-alloy steels

Code defines both a minimum service temperature and a minimum pressure test temperature

For PWHT vessels, the service temperature and pressure test temperatures are equal

For As-welded vessels, the pressure test temperature is 30°F higher than the service temperature for most vessels

At metal temperatures below the minimum service temperature, pressure applied to the vessel must be less than 20% of the required test pressure

This will usually be an applied pressure of 25% of the design pressure

Materials Selection Requirements

Either impact test exemption curves or C_V impact testing are used to assure that steels are above their transition temperature

Requirements are similar to Division 1 requirement (see the note in the table that follows)

Division 1 impact test exemption curves have been revised more recently than those in Division 2, Code Paragraph AM-218

The recommended exemption curve materials assignment for Division 2 to be used in conjunction with code Figure AM-218.1 is shown in the following table

The recommendations included in the table are slightly more conservative than those included in Division 1, Code Paragraph UCS-66, especially at thickness less than 1½ inches

Recommended Exemption Curve Materials Assignment for Section VIII, Division 2 (Figure Am-218.1)

PLATE	FORGINGS	PIPE
Curve 1		
SA 36 (nonpressure containing attachments only, ≤ ¾ inch thick)		
Curve II		
SA 285	SA 105	SA 53
SA 515	SA 181	SA 106
SA 387 (annealed)	SA 366 and 182 (annealed)	SA 355 (annealed)
Curve III		
SA 516 if not normalized		
SA 387, Gr. 11 and 12[1]	SA 182 or SA 336, Gr. 11 and 12[1]	SA 335, Gr. P11 and 12[1]
Curve IV		
SA 387, Gr. 21 and 22[1]	SA 182 or SA336, Gr. F21 and 22[1]	SA 335, Gr. P21 and 22[1]
Curve V		
SA 516 normalized	SA 350, LF 1 and 2[2]	SA 333, Gr. 1 and 6[2]
SA 537l, Cl 1		

[1]Normalized and tempered.

[2]SA 350 LF 1 and 2 and SA 333 Grades 1 and 6 are acceptable for minimum design temperatures down to −50°F without additional impact testing.

Charpy V-Notch Impact Testing

Requirements are provided in Code Paragraphs AM-204 through AM-218

C_V impact tests are required

When steel is to be used at a combination of metal temperature and thickness below its exemption curve

For all carbon and low-alloy steels thicker than 2 inches regardless of minimum pressurizing temperature

The impact test requirements of Division 1, Code Figure UG-84.1 are recommended in place of Division 2, Code Table AM 211.1

Requirements are less conservative

Do not take into account the need for higher C_V energy with increasing thickness

Typical carbon steel selections to avoid brittle fracture in pressure vessels are listed in the following table:

Typical Carbon Steel Selection to Avoid Brittle Fracture in Pressure Vessels

MINIMUM PRESSURING TEMP., °F	SHELL AND HEAD PLATES		NOZZLE	
	1-IN. THICK	2-IN. THICK	FORGINGS	PIPE
−50 to 0	Normalized 516[2,4] or SA 537 Class 1	Normalized[1,2,4] SA 516 or SA 537 Class 1	SA 350 LF2	SA 333
0 to 30	Normalized SA 516 or impact tested, As-rolled SA 516	Normalized SA 156	SA 105[(3,4)]	SA 106[3,4]
30 to 60	As-rolled SA 516 or impact tested SA 285	Normalized SA 516	SA 105[3,4]	SA 106[3,4]
Warmer than 60	SA 285 or As-rolled SA 516	Normalized SA 516	SA 105	SA 106

[1]The SA 516 specification requires plates 1½ inches and thicker to be normalized.

[2]Plates may require impact testing; see Code, Division 1, Figure UCS-66.

[3]Forgings and pipe may require impact testing; see Code, Division 1, Figure UCS-66.

[4]Impact testing of the heat-affected zone (HAZ) and deposited weld metal is required on the Welding Procedure Qualification Test Plate when impact testing is required on the parent metal.

Auto Refrigeration

Defined as the temperature that the contents of the vessel would reach if the vessel is depressured to 40% of its MAWP

If the temperature of auto refrigeration is less than 20°F, then the vessel should be treated as subject to auto refrigeration and this used as a design basis to avoid brittle fracture

Should be considered when selecting steels

In some liquid services, such as LPG, a leak could do the following:

Reduce the pressure

Cause a drop in temperature

Cause the liquid to boil off

Steel selection for pressure vessels subject to auto refrigeration

Vessels subject to auto refrigeration require additional considerations:

1. Steels from Code Figure UCS-66, curve D should be used; typically, carbon steel plate should be normalized SA 516; forgings may be SA 350, grade LF 2, and pipe may be SA 333, Grades 1 or 6; these steels have good inherent toughness
2. Impact testing is not required for auto refrigeration, unless already required at the normal design temperature; SA 350 and SA 333 materials are, however, impact tested in accordance with their respective specifications

Auto refrigeration is not considered equivalent to a cold design or operating temperature because of the lowered pressure; therefore, the recommended safeguards against brittle fracture are not as stringent as for a cold operating temperature; the use of SA 516 steel, and equivalent forging and piping grades, should by itself provide ample resistance to brittle fracture during auto refrigeration

Impact testing is not required for auto refrigeration, unless it is required for a cold design temperature without considering auto refrigeration

► GUIDELINES FOR PREVENTING BRITTLE FRACTURE IN EXISTING EQUIPMENT

General Considerations

The general rule is to limit a pressure vessel to less than 40% of its MAWP any time the vessel metal temperature is below the minimum pressurizing temperature

Be sure there are no other significant stresses such as those resulting from weight and thermal expansion

For vessels without a specified minimum pressurizing temperature, one can be established using the following guidelines:

The minimum pressurizing temperature (MPT) can be established from the following:

- Knowledge of the steel types
- Thickness of the vessel
- Minimum temperature for hydrotest, startup, or operation

Determining MPTs

Guidelines for Existing Vessels

Note that *minimum pressurizing temperature (MPT)* is the same as *minimum design metal temperature (MDMT)*

Carbon steel and low-alloy steel vessels less than 6 inches in diameter can establish an MPT from Code Figure UCS-66

Materials of construction and vessel thickness

Vessels made with materials shown on Code Figure UCS-66, the MPT, can be established directly from the curves if component thicknesses are known

The MPT is determined by the intersection of component thickness with the appropriate curve on Code Figure UCS-66

Consider only welded parts such as shells, heads, channels, and integrally reinforced nozzles for MPT

The component with the highest MPT sets the MPT for the vessel

Reinforcing pad nozzles or small nozzles without reinforcement does not need to be considered for an MPT unless the risks of brittle fracture warrant extra precaution

Fractures of vessels with reinforcing pad nozzles generally occur in the shell plate at the reinforcing pad-to-shell fillet weld

Integrally reinforced nozzles should be considered for MPT

Guidelines Clarifying the Use of Code Figure UCS-66 to Establish MPT

1. The thickness of vessel components refers to the thickness at a weld
2. An MPT does not need to be established for nonwelded parts like flanges or heat exchanger channel covers
3. To use the curves for normalized material, vessel records must indicate normalized material was used
4. For P-1 carbon steel vessels that were stress-relieved but were not required to be stress-relieved by ASME code, the MPT may be 30°F lower than that given by the exemption curve on Code Figure UCS-66; this is consistent with Code Paragraph UCS-68(c); normally, carbon steel vessels 1¼ inches and less in thickness are not required to be stress-relieved by code rules; refer to

code table UCS-23 to determine whether a steel is P-1 steel

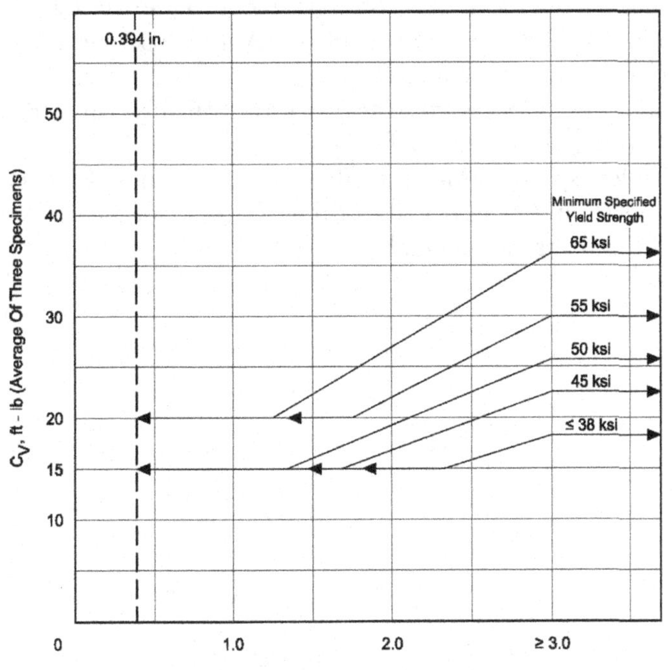

General Notes :

(a) Interpolation Between Yield Strengths Shown Is Permitted.

(b) The Minimum Impact Energy For One Specimen Shall Not Be Less Than 2/3 Of The Average Energy Required For Three Specimens.

(c) Materials Produced And Impact Tested In Accordance With SA-320, SA-333, SA-334, SA-350, SA-352 And SA-420(See Table UG-84.3) Do Not Have To Satisfy These Energy Values. They Are Acceptable For Use At Minimum Design Metal Temperature Not Colder Than The Test Temperature When The Energy Values Required By The Applicable Specification Are Satisfied.

(d) For Materials Having A Specified Minimum Tensile Strength Of 95 ksi Or More, See UG -84(c)(4)(b).

5. All grades of SA 285 and SA 515 steels thicker than ¾ inch should be assigned to curve A rather than curve B

Guidelines for Existing Vessels with Obsolete Steel Specifications

All steels, except the code case steels, should be assigned to curve A or Code Figure UCS-66, unless sufficient documentation is available for assignment to a lower curve

Code Case 1256 is equivalent to SA 442 (use curve B; if normalized, use curve D)

Code Case 1280 is equivalent to SA 516 (use curve B; if normalized, use curve D)

Obsolete specifications for tubes, pipes, forgings, and castings should be assigned to curve A unless specific data to the contrary are available

Mechanical Design of Pressure Vessels

4

▶ OVERVIEW

This chapter discusses the following topics:

Responsibilities of designing and fabricating pressure vessels
Design basis
Details or mechanical design
An overview of structural design
A quick reference guide to ASME Code, Section VIII, Division 1

▶ GENERAL CONSIDERATIONS

Most pressure vessels have the following features:

Cylindrical shells with elliptical or hemispherical heads
Simple to design, fabricate, and install in either vertical or horizontal position
Design is governed
Internal pressure
Few are designed to operate below atmospheric pressure
Most pressure vessels are designed to the rules of the ASME Code, Section VIII, Division 1
- Emphasized in this section
- Information in this section
  - Not intended to be a substitute for the code but rather to clarify the requirements of and to save time in using the code

▸ OWNER'S, USER'S, AND MANUFACTURER'S RESPONSIBILITIES

ASME Code, Section VIII

Assigns certain responsibilities to the owner/user, manufacturer, and authorized inspectors

When all discharged properly, provides for the design and construction of pressure vessels that will be safe to operate under the intended service condition

Provides minimum requirements for safe containment of design pressure at the design temperature

Does not give direct consideration to service conditions that can affect the performance of a pressure vessel

Owner/user is solely responsible for specifying requirements that exceed the code minimums depending on service conditions

This responsibility is a code requirement placed on the owner/user beyond the explicit rules of the code

ASME Code, Section VIII, Division 1

Owner's and User's Responsibilities

Code Paragraph U-2 defines the owner's/user's (company's) and manufacturer's responsibilities:

Must "establish the design requirements for pressure vessels, taking into consideration factors associated with normal operation and other conditions such as startup and shutdown

Code recognizes that all of its design rules and construction details may not be appropriate for all service conditions and thus makes the owner/user responsible

Task can be delegated to a "designated agent," such as the engineering contractor for a project, but the final responsibility for safe operation still resides with the owner/user

The need for corrosion allowance and Post Weld Heat Treatment (PWHT) beyond the requirements of the code are specifically mentioned in Code Paragraph U-2; most companies fulfill their responsibilities by specifying to the manufacturer the requirements for corrosion allowance and PWHT for each pressure vessel

Additional requirements that exceed the minimum code requirements are often imposed by the petroleum industry

It is important to increase the reliability and safety

Additional expense is compensated for by lower maintenance costs and reduced loss of production resulting from unscheduled shutdowns for repair

Most common requirements include the following:

1. Double-V butt welds or equivalent required for all girth, longitudinal, and head-to-head welds
2. Full penetration welds required for all nozzle and flange welds
3. Integrally reinforced nozzles required for shell components with a thickness greater than 2 inches and for operating temperature above 650°F
4. Assured resistance to failure by brittle fracture during startup and shutdown by requiring the materials of construction to have adequate C_V impact toughness at the minimum design metal temperature (MDMT)
5. Seismic and wind practices that exceed ANSI standards

Manufacturer's Responsibilities

Code Paragraph U-2 requires the manufacturer of a pressure vessel to do the following:

Hold a "Certificate of Authorization" from ASME

Satisfy an authorized inspector that all applicable requirements of the code have been met before applying the "U" stamp to the vessel's nameplate

Be responsible for the calculations required to determine the minimum thickness of all components of the vessel, regardless of mechanical design input from the company

Prepare a "Manufacturer's Data Report" that contains the following:

Design temperature and Maximum Allowable Working Pressure (MAWP)

Materials of construction

Corrosion allowance

Thickness of major components

Non-destructive examinations

Joint efficiencies

PWHT

Data report must be signed by the manufacturer and the authorized inspector which serves as proof that legal requirements have been met

Authorized Inspectors' Responsibilities

Receive a commission from the National Board of Pressure Vessel Inspectors to inspect pressure vessels for code compliance

Are usually employed by an insurance carrier that serves as the manufacturer's inspection agency

Cannot be directly employed by the manufacturer

Authorized inspectors are required to do the following:

Monitor manufacturer's quality control and Non Destructive Examination (NDE)

Verify that all required calculations have been made

Make all other inspections necessary to certify that the pressure vessel has been designed and fabricated according to all applicable rules of the code

Authorized inspectors are not required to do the following:

Verify the accuracy of the manufacturer's calculations

Verify compliance of NDE with code acceptance standards

The authorized inspector should not be relied on to do the following:

Verify that the owner's/user's (company's) additional specified requirements have been met

Verify that the pressure vessel is satisfactory for its intended service conditions

The owner's/user's (company's) pressure vessel engineers and inspectors, or contractors acting for the owner/user (company) should do the following:

Be involved with the design and construction

Assure that the owner/user (company) requirements are met

It is recommended that owner/user (company) engineers and inspectors make their own verification that code rules are complied with whenever design reviews of shop inspections are made.

ASME Code, Section VIII, Division 2

Owner's and User's Responsibilities

More explicit than Division 1

Contained in Code Article G-3

Required to prepare a "User's Design Specification" that gives "The intended operating conditions in such detail as to constitute an adequate basis for selecting materials and designing, fabricating and inspecting the vessel"

Significant differences from Division 1 are as follows:

Must indicate if a *fatigue analysis* should be made for cyclic pressure or temperature operation

Must provide sufficiently detailed information regarding the cyclic conditions to make the fatigue analysis

A fatigue analysis can complicate the design of a pressure vessel, but the owner/user may exempt the vessel from a fatigue analysis based on the successful operation of similar equipment under similar conditions

Manufacturer's Responsibilities

Design and construct the vessel to meet the conditions in the "User's Design Specification"

Comply with all applicable requirements of the code

Division 2 makes the manufacturer responsible for establishing the design requirements to meet the service conditions specified by the owner/user, whereas Division 1 leaves this up to the owner/user

A "Manufacturer's Design Report" must be prepared

- Includes all calculations necessary to show that the design, as shown on the fabrication drawings, complies with the code requirements
- Meets the conditions of the "User's Design Specification"
- Analysis of local primary membrane stresses, discontinuity stresses, and secondary (thermal) stresses should be included whenever they affect the design of a component of the vessel
- Fatigue analysis should be included if required by the "User's Design Specification"

Authorized Inspector's Responsibilities

Must review documentation concerning the following:

- Materials certification
- NDE
- User's design specification
- Manufacturer's data report

Verify that code requirements have been met before applying the "U-2" stamp to the nameplate

Not responsible for the following:

- Verifying the completeness or correctness of the design calculations
- Determining that the service conditions in the "User's Design Specification" have been appropriately addressed in the manufacturer's design

Division 2 vessels normally require greater involvement of user/owner (company) staff, because the accuracy of design and integrity of fabrication are critically important to the reliable performance of the vessel.

DETERMINING DESIGN CONDITIONS

General Considerations

The following design conditions must be established before actual design begins:

- Design pressure and temperature
- Wind and earthquake loads
- Corrosion allowance
- External and internal loads

Pressure and Temperature

Often govern the mechanical design

ASME code requires that a pressure vessel be designed for the "most severe condition of pressure and temperature expected in normal operation"

Pressure is directly entered into the code equation for design calculations

Temperature indirectly influences the design through its effect on the maximum allowable design stress for the materials of construction

Wind and Earthquake Loads

Specify the location where the vessel will be installed because this must be considered

Greatest effect is usually on the support and anchoring of a vessel, with no effect on the design of pressure-containing components, unless the shell is relatively thin with respect to its diameter or the contents of the vessel are relatively heavy

Corrosion Allowance

Usually added to the minimum required thickness for each component, or a corrosion-resistant cladding is provided to protect the vessel from the service environment

External Loads

Major loads are from piping connections to nozzles or from structural attachments to the shell and heads, such as support clips for platform and ladders

These resulting loads applied to a vessel have a negligible effect on the design of nozzles and the reinforcement of the openings in the shell

Design Pressure: Overview

The following types of pressure must be considered in the design of a pressure vessel:

- Operating pressure
- Design pressure
- Maximum allowable working pressure (MAWP)

Figure 4.1 is a schematic illustrating a typical pressure vessel with the following:

- Pressures that must be considered
- Relationships among these pressures

These pressures are also illustrated in Figure 4.2

The terminology used is consistent with ASME Code, Section VIII, Divisions 1 and 2

Design Pressure: ASME Code, Section VIII, Division 1

This section discusses the following:

- The various pressures in more detail
- How they are determined
- How they are used for the design of a pressure vessel

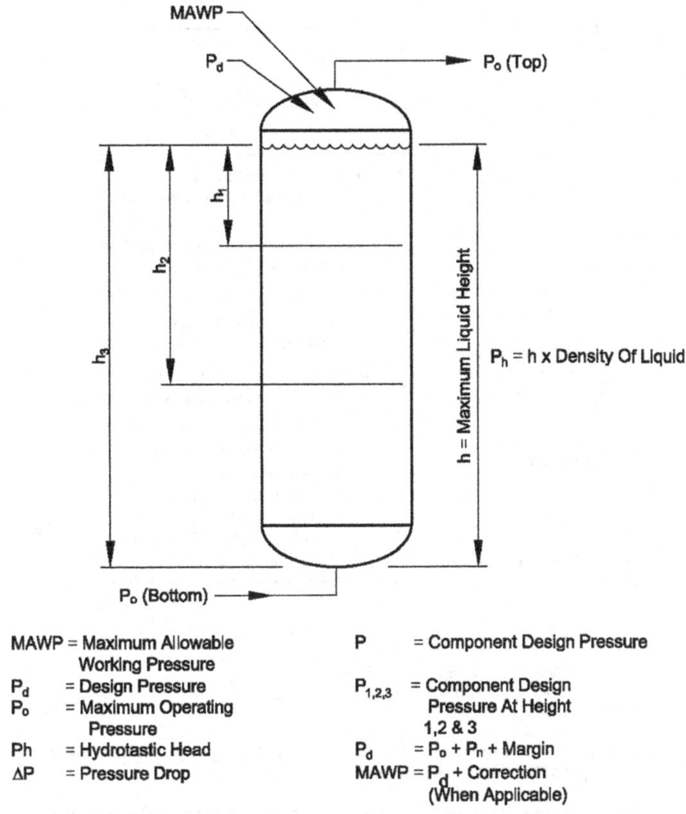

MAWP = Maximum Allowable Working Pressure
P_d = Design Pressure
P_o = Maximum Operating Pressure
Ph = Hydrotastic Head
ΔP = Pressure Drop
P = Component Design Pressure
$P_{1,2,3}$ = Component Design Pressure At Height 1,2 & 3
P_d = P_o + P_h + Margin
MAWP = P_d + Correction (When Applicable)

Figure 4.1 Representative pressure vessel with definitions of pressure considered in design.

Maximum Allowable Working Pressure (MAWP)

Required to be displayed on a pressure vessel's nameplate

Defined in the code as "the maximum pressure permissible at the top of the vessel in its normal operating position at the (design) temperature"

Is not the same as design pressure (P_d), which provides the basis for the design of the vessel

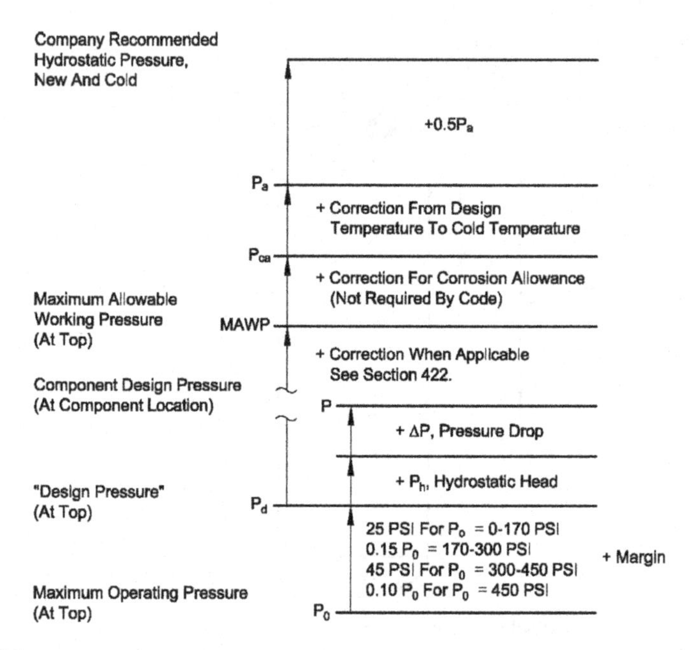

Notes

1. Correction = $\left[\dfrac{\text{Maximum Allowable Cold Stress}}{\text{Maximum Allowable Stress At Design Temperature}}\right] \times P_{ca}$
2. MAWP Is The Maximum Pressure Allowed In Service, At Design Temperature. MAWP May Exceed P_d, As Explained In Section 422.
3. Component Design Pressure Is That Used In Code Equations To Size Components.
4. $\Delta P = P_{o\,(Bottom)} - P_{o\,(Top)}$ (Negative For Downflow Vessels).
5. ADD P_h When Applicable (Liquid Filled Vessels).
6. Margins Shown Are Normal Company Recommendations.
7. Obtain P_o From Process Designer.

Figure 4.2 Schematic Illustration for Determining Hdrostatic Text Pressure.

Is determined from the design of the vessel as described next and is not used for the design

Design Pressure and Operating Pressure

Maximum operating pressure (P_0)

Specified by process design engineers

Maximum internal pressure that will occur under normal process conditions

Design pressure (P_d)

Determined by adding a margin to P_0 to allow for pressure surges above P_0 (without lifting the pressure safety valve)

P_0 should be increased by the following minimum margins to obtain P_d

P_0. PSIG	MARGIN
0-170	25 psi
170-300	0.15 P_0
300-450	45 psi
450+	0.10 P_0

These margins may have to be increased if the designers anticipate higher pressure surges

Design Pressure and Pressure Used for the Design of Individual Components

P_d establishes the basis for the design of a vessel, but it is not always the pressure used for designing components of the vessel

Pressure at the bottom of a vessel containing a liquid is higher than the pressure at the top because of the hydrostatic head (P_h)

P_h from the top of the liquid to the component being designed must be added to P_d to establish the component design pressure (P), which is used in the code design calculations for that component ($P = P_d + P_h$ for liquids)

If a vessel only contains gas then (P_h) is negligible ($P = P_d + 0$)

Only one value for P is determined for the bottom of the vessel, which is used for the design of all components of the vessel from top to bottom

May be advantageous to design tall vessels with low internal pressures for actual hydrostatic heads that exist at various levels in the vessel

Pressure Drop

If the design indicates a large pressure (ΔP) through the vessel, it should be added to establish the component design pressure (P):

$$P = P_d + P_h + \Delta P \tag{1}$$

Calculation of Maximum Allowable Working Pressure (MAWP)

The actual thickness of the various vessel components will be greater than the thickness calculated using P

More economic to purchase the next thicker commercial size of plate, pipe, or ANSI B16.5 flange than to have component fabricated to the exact thickness required

The MAWP permitted will be somewhat higher than the design pressure (P_d)

Code allows calculating an MAWP based on this extra thickness, adjusted for P_h, for each vessel component, using the lowest MAWP for any component as the MAWP for the vessel

If the MAWP is not calculated for the actual component in this described manner, P_d must be used for the MAWP on the nameplate

When the design pressure (P_d) is used for the MAWP on the nameplate, the entire thickness should be added to the corrosion allowance for each component of the vessel

The MAWP of a vessel should not be limited by the MAWP of a minor component, such as a flange or nozzle

If an ANSI B16.5 flange has a lower pressure rating than the MAWP for the shell and head components, the flange should be upgraded to the next higher class

Upgrading can cause complications if the associated piping class calls for lower pressure flanges, and a nonstandard flange must be added to the pipe

These factors must be evaluated for each circumstance

Design Pressure: ASME Code, Section VIII, Division 2

Pressures are similar to those discussed for Division 1, but terminology differs

Design Pressure (P_d)

Maximum pressure permissible at the top of the vessel at the design temperature is displayed as such on the vessel's nameplate

May not be the pressure used to design the components

(P_h) must be added to (P_d) to determine the pressure (P) used for the design of each component ($P = P_d + P_h$)

Division 2 requires that the coincident pressure ($P_d + P_h$) at any point should be used for the design of a component

Division 2 does not refer to an MAWP

Each component of a pressure vessel will usually be fabricated from custom-produced materials close to the minimum thickness plus corrosion allowance, with the exception of ANSI B16.5 flanges

Therefore, no benefit will be obtained by calculating an MAWP for the actual thickness

External Pressure

Most vessels are designed to contain positive internal pressure

Possible to develop a partial vacuum inside the vessel

- During steam-out cleaning
- When draining liquids
- During abnormal process conditions

Therefore, it is general practice to check all vessels for their resistance to collapse under 7.5 psi external pressure at 450°F

Normally will affect only large-diameter vessels that are designed for low internal pressures

If the minimum required thickness for the internal pressure does not provide adequate resistance to collapse under a partial vacuum, it is usually more economic to add stiffening rings to the shell than to increase thickness

If a vessel design exceeds the required minimum external pressure rating of 7.5 psi, the vessel should be stamped for the higher pressure, up to 15 psi

Design Temperature: ASME Code, Section VIII, Division 1

Many pressure vessels are designed

Operate at high temperatures

Required to be under pressure at ambient temperature (startup or shutdown conditions)

Some vessels are required to operate at temperatures below ambient temperatures, whereas others can be subjected to "auto refrigeration" resulting from operational upsets or gas leaks

Both Divisions 1 and 2 require the nameplate of a vessel to display both a maximum temperature and a minimum temperature

Maximum Temperature

Controls the vessel design by establishing the maximum allowable design stresses for the selected materials of construction

Assures that the stresses developed in a pressure vessel at the MAWP will not cause failure by ductile bursting or gross yielding during continuous operation at the maximum temperature

Minimum Design Temperature

Does not directly affect the design

Can affect the selection of the materials

Materials must have sufficient toughness at the minimum design temperature to prevent failure by brittle fracture during startup and shutdown

Division 1 defines both maximum and minimum design temperatures, which establish a temperature range over which the vessel can be operated at the MAWP

Maximum Design Temperature

As defined in Section 1, "Shall not be less than the mean metal temperature (through the thickness) expected under operating conditions"

Displayed on a vessel's nameplate as the temperature for which the MAWP is determined

It is the maximum metal temperature permitted at the MAWP

Establishing the Maximum Design Temperature

Based on the normal maximum operating temperature obtained from process design

Standard practice is to add a margin of at least 25°F above the maximum operating temperature to assure safe operation in the event of a minor upset

When the maximum operating temperature is below 650°F, it may be desirable to increase the maximum design temperature to 650°F to provide greater flexibility

Possible without altering the design of the vessel (thickness), because the maximum allowable design stress for most materials is the same for any design temperature up to 650°F

However, the pressure ratings for ANSI B16.5 flanges does change at temperatures below 650°F, and increasing the temperature to 650°F may necessitate upgrading the flanges to the next higher class

A maximum design temperature above 650°F significantly affects design because much lower stresses are permitted by the code

Minimum Design Metal Temperature (MDMT)

Defined in Section 1 as "the lowest (temperature) expected in service"

Same as the minimum pressurizing temperature (MPT) previously discussed in the Materials section

For most vessels, it is the lowest metal temperature permitted at the MAWP

Current practice is to limit the operating pressure to 40% of the MAWP at temperatures below MDMT

The minimum design temperature should never be above 50°F, unless the circumstances are discussed with an experienced pressure vessel engineer

Must assure resistance to brittle fracture at the minimum design temperature; this is accomplished by

- Selecting materials that are known to have adequate toughness at the minimum design temperature or
- Requiring C_V impact testing in order to establish that adequate toughness exists

Operation below Ambient Temperature

Vessels that continuously operate at temperatures below normal ambient temperatures are considered to be in "critical" service because of the risk of brittle fracture

MDMTs are based on a vessel's normal minimum operating temperatures

Minimum design temperatures are usually designated at 10°F below the minimum operating temperatures

Startup and Shutdown at Ambient Temperature

Vessels located in severe winter climates may be required to be at maximum operating pressure at low ambient temperatures during transient startup or shutdown

MDMT should be the lowest ambient winter temperature that would be expected during severe winter conditions

Operating pressure must not exceed 40% of the MAWP until the vessel is warmed up to the MDMT

Transient startup and shutdown conditions are not considered to be critical service, unlike continuous operating at low temperatures

Restricted Startup and Shutdown

Most vessels in locations with a mild winter climate will not require a special materials section or C_V impact testing to assure adequate resistance to brittle fracture during startup and shutdown

MDMT for each component can be established by determining the minimum temperature permitted by the code for the material of construction and nominal thickness

The maximum temperature determined in this manner would be designated as the MDMT for the vessel

Operating pressure during startup and shutdown should not be allowed to exceed 40% of the MAWP at lower temperatures, but this should not place prohibitive restrictions upon startup and shutdown procedures

Auto Refrigeration

Defined as the temperature that the contents of a vessel would reach if the vessel is de-pressured to 40% of the MAWP

Not considered a critical service condition because the loss of pressure that causes the contents of a vessel to auto refrigerate also reduces the operating stresses during auto refrigeration

Cooling is likely to be highly localized at the source of a leak or will lag behind the temperature of the vessel's contents during a general system loss of pressure

Vessels should be fabricated from SA 516 normalized or SA 537 plate (with equivalent pipe or forging grades)
- Offer good toughness at low temperatures
- Provide adequate resistance to brittle fracture

C_V impact testing is not mandatory for the auto refrigeration temperature

Hydrotesting

Hydrotesting a pressure should never be performed at an ambient temperature below the minimum design metal temperature (MDMT)

The high stresses developed during a hydrotest would make the vessel especially vulnerable to failure by brittle fracture

Design Temperature: ASME Code, Section VIII, Division 2

Maximum Temperature

Division 2 requires that the following for the design temperature:
- "Shall be based upon the actual metal temperature expected under operating conditions"
- Should be determined from the maximum operating temperature in the same manner as described for Division 1

Design temperature
- Maximum temperature permitted with the design pressure
- Displayed as such on the nameplate

Minimum Temperature

A minimum permissible temperature is required to be displayed on the nameplate of a Division 2 vessel

Unlike the nameplate for a Division 1 vessel, a corresponding pressure is not displayed

The pressure during startup or shutdown is not permitted to exceed 25% of the design pressure (defined in the code as 20% of hydrotest pressure) at temperatures below the minimum permissible temperature

The minimum permissible temperature can be determined in the same manner used to determine this factor for a Division 1 vessel

The curves concerning C_V impact test requirements for the various materials of construction tend to require C_V impact testing at slightly higher temperatures than in Division 1

This is probably related to the greater toughness required to resist brittle fracture at the higher design stresses in Division 2

Wind and Earthquake Design

Wind and earthquake loadings are specific to the geographic location where the vessel is installed

Both wind and earthquake loads create overturning moments that develop longitudinal stresses in the shell of vertical vessels

The weight of the internal contents of a vessel must do the following:

Amplify the overturning moment resulting from earthquake loading

Be taken into consideration when calculating the longitudinal stresses

Not necessary to design for the simultaneous occurrence of maximum wind and earthquake loads

Longitudinal stresses developed in the vessel's shell by the wind and earthquake loads must be added to the longitudinal stresses attributable to the internal pressure

The longitudinal stress attributable to the internal pressure is normally one half of the hoop stress, which is the maximum principal stress that governs the design of the vessel for internal pressure

The combined stresses for wind or earthquake and internal pressure allowed exceed by 50% the maximum allowable design stress for the material of construction given in the code

Higher stresses are permitted because the vessel is only intermittently subjected to the severe wind and earthquake loads

Consequently, the wind and earthquake loads will not usually affect the design of a pressure vessel shell
The major exception would be a vessel designed for low internal pressure with relatively thin shells and a high weight of internal contents
Wind and earthquake loads can have a significant effect on the design of the support and anchoring for a vertical vessel

Corrosion Allowance

Internal service conditions a pressure is exposed to during operation can cause materials to corrode
Corrosion allowance is normally added to the calculated minimum thickness required for each component of the vessel
Code assigns to the owner/user of the vessel the responsibility for specifying the corrosion allowance
This is necessary to prevent corrosion from reducing the thickness below the required minimums during operation
An alternative, especially when the process environment would result in a very high corrosion rate, is to do the following:
Employ a corrosion-resistant cladding
Apply a corrosion-resistant coating
External corrosion is rarely significant, and thus an external corrosion allowance is not usually necessary
Usually, either pressure vessels operate at high temperatures that prevent the condensation of moisture or weather shielding is provided

External and Internal Loads

External Loads

External loads applied to a pressure vessel are usually of a local nature

Stresses developed in the vessel's shell by local external loads must be added to the stresses attributable to the internal pressure, normally either local primary membrane stresses or bending stresses

Therefore, the total stress obtained by adding the stresses developed by local external loads to those attributable to the internal pressure is permitted to reach 1.5 times the maximum allowable design stress given for the material of construction in the code

Consequently, only relatively high external loads are likely to affect the design of a vessel

Piping Connections

Stresses can be developed in a pressure vessel shell because of forces and moments that result from piping connections to nozzles

The magnitude of these forces and moments applied to a vessel are insignificant and need not be considered for the design of a vessel

This is especially true for vessels designed to Division 1 standards, where the required safety factor of 4 is sufficient to compensate for these small loads without detailed analysis

Heavy equipment attached to nozzles (valves and bridles) should be supported to minimize the external loads acting on the nozzle

Forces and moments attributable to piping connections can be calculated using the computer program CAESAR, if it is suspected that they are high enough to affect the design of the vessel

Structural Attachments

Platforms and ladders are frequently supported by direct attachment to a pressure vessel

Normal practice is to provide a sufficient number of clips welded to the vessel's shell such that the local load transmitted to the shell by any one slip is not great enough to affect the design of the vessel

Not practical for vessels with low design pressure and relatively thin wall

External load can be distributed over a larger area to reduce the stresses developed in the shell by providing a reinforcing pad on the shell for attachment of the clip

Not advisable to use a reinforcing pad if the operating temperature of the vessel will exceed 450°F; other design approaches should be investigated

Lifting Lugs

Must be provided for moving and erecting a vessel

The location of these lugs and the loads that will be applied to them depend on how the vessel will be moved and erected

Details worked out between vessel fabricator and construction contractor responsible for erection

Rigging diagram should be provided to the fabrication by the construction contractor

Static loads on the lugs resulting from the weight of the vessel are usually multiplied by a factor, depending on the construction contractor's lifting procedure and experience

There are no code criteria for maximum allowable stresses that can be used for design of the lifting lugs

They are designed to prevent damage to the vessel during installation only, and they have no effect on the integrity and reliability of the vessel during operation

Internal Loads

Pressure vessels normally contain various internal components that are attached directly to a vessel's shell, such as the following:

Distributor trays

Catalyst support grids

Baffles

De-mister pads

These internal components apply loads to the shell and thereby develop stresses that must be added to those resulting from the internal pressure

The weight of the internal components plus the weight of liquid or catalyst supported by the component must be considered

The pressure drop across the component will apply an additional load to the shell that must be considered separately from the influence of the pressure drop of the design pressure

The internal loads in a vertical vessel are downward, developing a compressive stress in the vessel's shell that counteracts the longitudinal tensile stress developed by the internal pressure

It is rare that internal loads affect the design of a vertical vessel

An exception could be encountered with an upflow vertical vessel, if a high-pressure drop occurs across an internal component; this would develop a tensile stress in the shell that would add to the longitudinal stress developed by the internal pressure

The weight of the internal contents of a vessel (internal components, catalyst, fluids, etc.) will affect the design of the vessel's support

Directly increasing the compressive stress

Indirectly by amplifying the overturning moment in the event of an earthquake

▸ MECHANICAL DESIGN

Overview

Use of the ASME Code

Vessels must be designed to the latest edition of the ASME Code, Section VIII, and addenda

This section only discusses Division 1 rules

For application of Division 2 rules, consult an experienced pressure vessel engineer

This section will help you understand the code and how to apply it

There are two approaches to designing and ordering a pressure vessel:

One way is to specify only the required service conditions, nozzle sizes, and orientations to the fabricator and rely on that person to determine all design details of the vessel construction

Another approach is for the owner to determine many of the construction details prior to ordering the vessel; this approach causes the owner more upfront design engineering

Section VIII covers all pressure vessels other than those required to be in accordance with Section I, III, or IV

Fired process tubular heaters are specifically excluded

Vessels excluded from Section VIII are the following:

Those whose internal design pressure does not exceed 15 psig with no limitation on size

Those with an inside diameter that does not exceed 6 inches, with no limitation on pressure

Piping versus Pressure Vessels

It is often difficult to decide if a unit being incorporated into a piping system should be classed as piping or as pressure vessel

No precise differentiation exists, and thus the classification must be left to "engineering judgment"

The unit should be called *piping* in the following conditions:

1. As a part of the piping system, its primary function is to *transport fluid* from one location to another within the system (header or manifold are examples)

   Special design features or accessories added to permit secondary functions would not change its classification (enlargement of any part or all of a header to provide a degree of pulsation dampening is an example)
2. The element under consideration is available from, and is classified by, recognized piping equipment suppliers as a *piping component* or accessory (strainers, filters, steam traps, expansion joints, and metering devices are examples)

Units that are normally constructed in accordance with the code should not be included in this category but should be classified as pressure vessels

Even if fabricated exclusively from pipe and fittings, a unit other than a commercial piping accessory should be classified as a *pressure vessel* in the following conditions:

1. Its primary purpose is not to transport fluid but to process fluids by distillation, heat exchange, separation, or removal of solids
2. Its primary function is to store under pressure

Pressure/Temperature Limitations

Figure 4.3 shows the design pressure and design temperature range for various design codes and standards

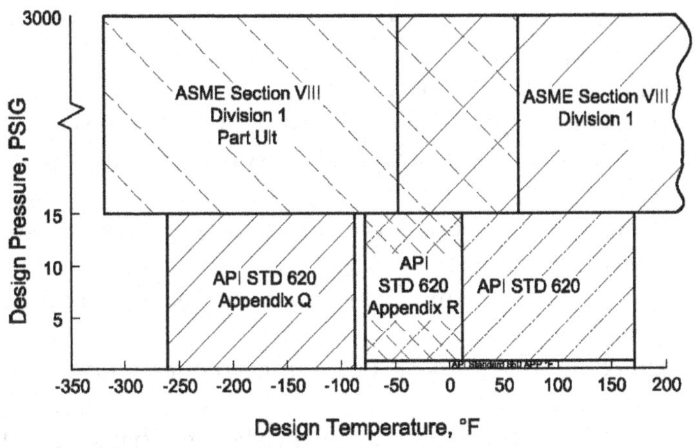

Figure 4.3 Scope of various design codes and standards.

Design Considerations

Code rules do the following:

- Provide minimum requirements for safety and service
- Hold owner/user and the designer/manufacturer responsible for meeting code rules as well as the service needs
- Require an ASME code stamp on a vessel
- Provide that new and cold, the vessel is good for the pressure and temperature indicated and recorded on the manufacturer's ASME Data Report
- Permit the legal operation of the vessel in a jurisdiction that has adopted the ASME code as part of the law

Because of the range of materials and potential services for pressure vessels constructed under Section VIII, it is understandable that the code must hold the owner/user and manufacturer responsible for constructing a pressure vessel for a specific service

The owner's/user's designated agent may be any of the following:

Engineer on the engineer's staff

Design agent specifically engaged by the owner/user

Manufacturer or designer of a system for a specific service

Organization that offers pressure vessels for sale or lease for a specific service

Engineer of the stamp holder that will manufacture the pressure vessel

Design conditions must include, but need not be limited to, the following:

1. The need for corrosion allowance beyond that specified by the rules, except for a limited number of vessels constructed of low-carbon steel containing air, steam, or water; corrosion allowance is not a requirement of the code
2. The contents defined with particular attention to toxic substances so that the mandatory requirements of the code can be met
3. The need determined for impact testing or base materials and deposited weld metal if the vessel will operate at low temperature
4. The need for post-weld heat treatment beyond code mandatory requirements that might be necessary for resistance to corrosion or because of the service environment
5. Proper selection of the materials for the service conditions
6. Nozzle locations and external piping reactions that will occur on the nozzles during normal operation
7. The installed orientation of the pressure vessel and the type of support required

8. For pressure vessels generating steam, the need for piping, valves, instrumentation, and fittings required for the service as set forth in Section I, Code Paragraphs PG-59 through PG-61
9. The number of openings and the size for the required safety relief devices
10. Selection of vessel heads: very high compressive stresses can exist in the knuckle region of torispherical heads under internal pressure; the magnitude of the stress is primarily a function of the knuckle radius, the diameter, and the diameter-to-thickness ratio
11. Openings in knuckles: caution should be exercised in locating openings or another attachment that may produce large stress concentrations in the region of the knuckle of torispherical heads or the shell-to-cone junction of the conical transitions
12. Structural, wind, and earthquake: the code does not have design rules for wind or earthquake loads; therefore, design for wind and earthquake loads should be in accordance with company design standards

Recommended Practices

Code rules establish minimum construction standards

Because of their wide scope, higher-quality standards are usually justified, except for the simplest of vessels in noncritical services

UM vessels (these vessels have a volume and pressure limit)

Utility pressure vessels, such as the air receivers supplied as part of a small noncritical air compressor package and small noncritical refrigeration package equipment

Areas where company specifications usually exceed code requirements are as follows:

Protection against brittle fracture

Types of welded joints nozzles
Limits on using reinforcing pads and torispherical heads
Limits on certain welding procedures
Radiography
Stress relief of cold-formed parts
Preheat and post-weld heat treatment requirements
Hydrostatic tests

Protection against Brittle Fracture

At low temperatures, strength is less a consideration than the increasing susceptibility to brittle fracture

Brittle fractures generally occur without warning at stresses below the yield strength of the material

Carbon and low-alloy steels are subject to brittle fracture at low temperatures and even at temperatures significantly above ambient, depending on the thickness, method of manufacture, and so on

Post-Weld Heat Treatment (PWHT) for Service Conditions

PWHT is performed to reduce residual stresses, hold dimensional tolerances, improve resistance to brittle fracture, reduce susceptibility to stress corrosion, or provide added safety in certain hazardous services

In addition to thickness requirements, the code requires PWHT for carbon or low-alloy vessels operating below minus 50°F, for unfired steam boilers at pressure above 50 psig, and for lethal-service vessels

Where service conditions might promote stress corrosion cracking, the need for PWHT should be investigated

Hydrostatic Tests

The purpose of hydrostatic testing is to reveal defects in the vessel's workmanship and to detect the presence of leaks

Most weld defects cannot be determined by the use of hydrostatic testing and must be determined by other forms of nondestructive testing

Code rules permit an initial hydrostatic test pressure to be set at 1.3 (1.5 prior to July Addenda 99) times the MAWP or the design pressure to be stamped on the vessel

May be adequate for vessels of simple configuration used in noncorrosive services

For large or other complex vessels, or where a significant corrosion allowance is provided, this pressure rating can result in operating the vessel at stresses *above* those induced by the hydrostatic test (after corrosion has reduced the thickness of the vessel wall)

Hydrostatic test pressure should be calculated to stress the full thickness, including corrosion allowance and cladding of the strongest Category A weld to 130% (earlier editions 150%) of the code allowable stress, at the test temperature, provided no component of the vessel is stressed above 90% of the minimum yield strength

The pressure should also be reduced as necessary to avoid overstressing Category A welds

Industry experience indicates that, for test purposes, an upper limit on stress equal to 90% of the minimum specified yield strength of the base metal can be used, provided no excessive stress raisers are present

Common stress raisers:

- Improperly reinforced openings
- Poorly designed transitions between shell sections and between shells and heads (especially in the case of cone-to-cylinder intersections)
- Torispherical heads

Designs for especially large or thick vessels or unusual details should be reviewed by a stress analyst, especially in cases where test stresses approach the yield strength

Hydrostatic testing must also be conducted above the minimum design metal temperature (to avoid brittle fracture)

If a certain Category A weld cannot be fully tested as outlined here without overstressing other parts, the following alterations should be made:

Heads and transition sections should be thickened

Category A welds that still cannot be fully tested should be 100% radiographed

Cautionary safety note: Hydrostatic or pneumatic pressure testing is dangerous and has caused many injuries. Testing, therefore, must be conducted by those trained to do so.

The following guidelines are important to observe:

1. Avoid being within the vicinity of any pressure test; let the authorized inspectors and those with direct responsibility for testing the vessel read the pressure gauges and check for leaks
2. Stay away from the vessel when it is being pressured
3. Never make a close-up visual examination for leaks, particularly when the vessel is being pressured

Authorized personnel can make leak checks, after the vessel has been de-pressured to design pressure, using safe methods

High-pressure leaks eject high-velocity jets of gas/liquid mixtures that can penetrate deep into the skin or eyes

People have been blinded by visual leak checking without following proper precautions:

1. Most fabricators locate the hydrostatic pump and its pressure gauge at some nominal distance away from the vessel during the testing procedure so they need not get too close to the vessel; stay away from the pump and its

gauge and tubing as well; sometimes fittings break off or the gas leaks, and anyone in the path of the leak or fittings may be injured
2. Catastrophic failure during pressure testing is a possibility; there is a common belief that because the liquid compressibility is low, the stored energy and the energy released during a vessel failure is low; on the contrary, significant energy in hydrostatic tests come from the compressibility of the water, the expansion of the vessel, energy stored in dissolved gases, and, most significantly, trapped air that may not have been vented
3. The energy of pneumatic testing is far greater than hydrotesting because of the high compressibility of gas

Avoid being within the vicinity of any vessel under pneumatic testing

Design Summary

A summary of the mechanical design procedure follows:

1. Determine the service, contents, and conditions for the vessel
2. Determine the design pressures and temperature
3. Divide the vessel into its code components; these include the shell, heads, transitions, and nozzles
4. Identify the weld efficiency for each weld in each component
5. Establish the material's allowable stresses at design temperature
6. Establish a corrosion allowance
7. Use code procedures to size components for all loads; code equations are summarized in Figure 4.4 for frequently used components under internal pressure
8. Review the appropriate design considerations and the need for higher-than-code provisions

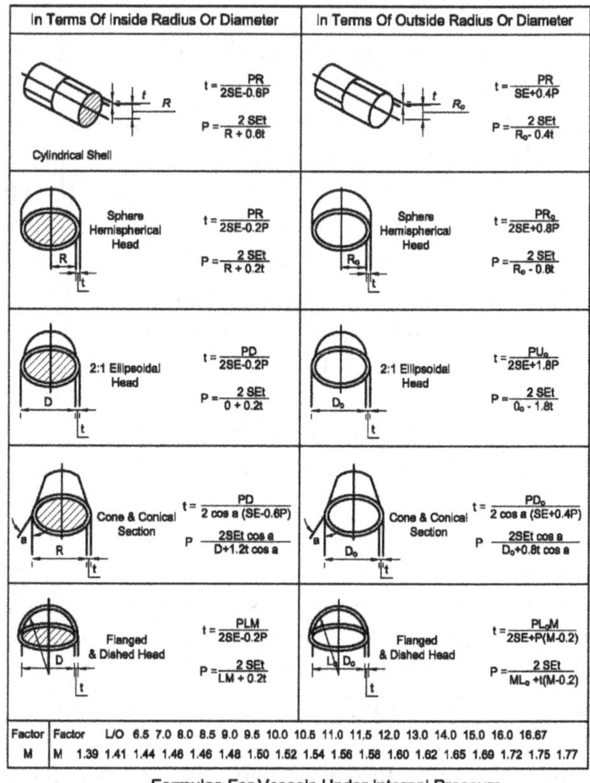

Notation

a = Half Apex Angle Of Cone, Deg.
D = Inside Diameter, Inches
D_o = Outside Diameter, Inches
E = Efficiency Of Welded Joints

L = Inside Crown Radius, Inches
L_o = Outside Crown Radius, Inches
M = Factor, See Table Above
P = Design Pressure or Maximum Allowable Pressure, psig

R = Inside Radius, Inches
R_o = Outside Radius, Inches
S = Stress Value Of Material, psi
t = Thickness, Inches

Figure 4.4 Equations for vessels under internal pressure.

9. Determine adequacy of external pressure; the procedure is complex but should be performed for all vessels subject to partial vacuum
10. Integrate and document the design

Design for Internal Pressure

General Considerations

Figure 4.4 is a summary of the equations for various components under internal pressure

Internal or external pressure produces a longitudinal seam stress that is two times larger than that on the circumferential seams

For this reason, the equations are shown only for longitudinal seams

However, if widely varying joint efficiencies for the circumferential and longitudinal seams are chosen or other stresses, such as structural stresses, come into play, then the circumferential seam may govern; this circumstance should be checked individually for each vessel under consideration

For a thick walled vessel where the wall thickness exceeds one half of the inside radius, or P exceeds 0.385 SE (S and E are defined in Figure 4.4), the equations given by Code Appendices 1 and 2 should be applied

Cylindrical Shell under Internal Pressure

AMSE Code, Section VIII, Division 1, Code Paragraph UG 27(c)(1) uses an approximate equation for determining the required thickness of a cylinder subjected to internal pressure:

$$t = \frac{PR}{SE - 0.6P} \quad (2)$$

Where

t = Required thickness of shell, inches
P = Internal design pressure, psi
S = Allowable stress, psi

R = Inside radius of cylinder, inches
E = Joint efficiency factor, dimensionless

Example

Vessel ID: 96 inches (fabrication ID)
Design Pressure P: 100 psig
Plate Material: SA 515-70
Design Temperature: 100°F
Corrosion Allowance: 0.125 inch (corroded ID = 96.250 inches/designing for retirement thickness)
All circumferential and longitudinal seams are double-welded butt joints and are spot radiographed.

Determine

The vessel is to be built per ASME Code, Section VIII, and Division 1.

Solution

1. From Table 1A of Section II, Part D, for SA-515 Grade 70 at temperatures up to 650°F, the stress value S = 17,500 psi.
2. From Table UW-12 of Section VIII, Division 1, for double-welded (Type 1) butt welds that are spot radiographed, the joint efficiency, E = 0.85.
3. The required wall thickness, t, is determined from the following:

$$t = \frac{PR}{SE - 0.6P} = \frac{(100)(48.125)}{(17,500)(0.85) - (0.6)(100)}$$
$$= 0.325\ inch + 0.125(\text{Corrosion Allowance})$$
$$= 0.450\ inch$$

Use a 0.5-inch plate.

Spherical Shell and Hemispherical Heads under Internal Pressure

Code Paragraphs UG-27(d) and UG-32(f) use the following approximate equation:

$$t = \frac{PR}{2SE - 0.2P} \tag{3}$$

The calculated minimum thickness of formed heads is rounded up to standard plate because of the thinning that occurs in portions of the head during forming (normally 1/16 of an inch)

The calculated value should be the minimum thickness at any point on the head; using the same vessel example,

$$t = \frac{(100)(48.125)}{(2)(17,500)(0.85) - (0.2)(100)} = 0.162\ inch$$

The calculated thickness should be increased by corrosion allowance:

$$t = 0.162 + 0.125 = 0.287\ inch$$

Semi-ellipsoidal Heads under Internal Pressure

Code Paragraph UG-32(d) uses the following approximate equation for 2:1 semi-ellipsoidal heads:

$$t = \frac{PD}{2SE - 0.2P} \tag{4}$$

Where

D = Inside diameter of the head

Code Paragraph UG-12(d) states that "seamless vessel sections or heads shall be considered equivalent to welded parts of the same geometry in which all Category A welds are type no. 1

E = 1 when spot radiography requirements of UW-11(a)(5)(b) are met

E = 0.85 when spot radiography requirements of UW-11(a)(5)(b) are not met or when Category A or B welds connecting seamless vessel sections or heads are type nos. 3, 4, 5, or 6 or Table UW-12"

Using the same example, with E = 1.0, then head thickness is as follows:

$$t = \frac{(100)(96.25)}{(2)(17,500)(1) - (0.2)(100)} = 0.275 + 0.125$$

$$= .400\ inch$$

Torispherical Heads under Internal Pressure

Code Paragraph UG-32(e) uses the following approximate equation for torispherical heads where the knuckle radius is 6% of the inside crown radius:

$$t = \frac{0.885\ PL}{SE - 0.1\ P} \quad (5)$$

Where

L = Inside crown radius

Using the same example, where L = 96 inches, E = 1.0 (seamless head), the head thickness is as follows:

$$t = \frac{(0.885)(100)(96)}{(17,500)(1) - 0.1(100)} = 0.195 + 0.125$$

$$= 0.320\ inch$$

Design for External Pressure

General Procedure

Figure 4.5 shows a flowchart for the general procedure for determining the maximum allowable external pressure on a vessel

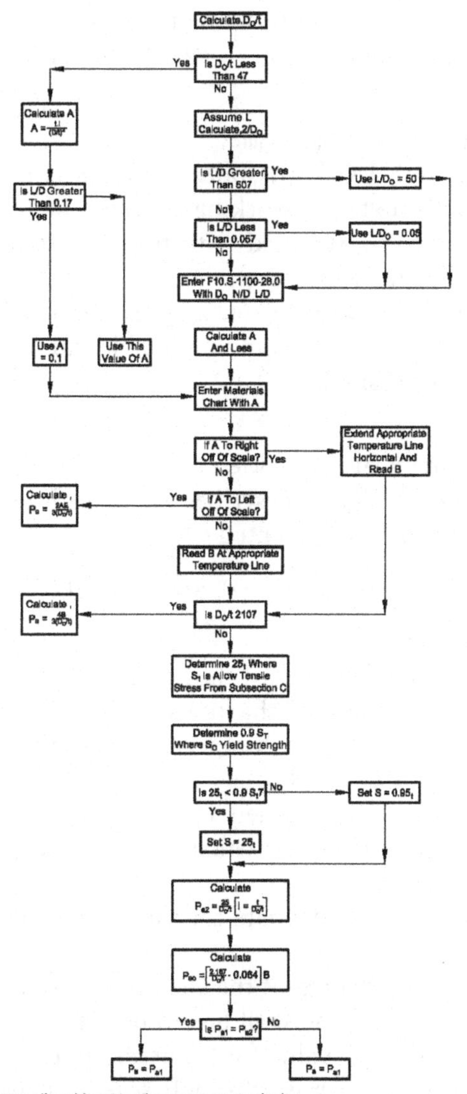

Figure 4.5 Maximum allowable external pressure on a cylinder.

The flowchart is based on the requirements of Code Paragraph UG-28 and Section II, Part D, Subpart 3

Cylindrical Shell under External Pressure

Code Paragraph UG-28(f) states that vessels intended for service under external working pressures of 15 psi or less and which to be stamped with the code symbol or are designed for the following:

A maximum allowable external pressure of 15 psi, or

25% more than the maximum possible external pressure, whichever is smaller

Cylindrical Shell under External Pressure

Code Paragraph UG-28 describes the procedure in detail

Two alternatives are presented

$$D_0/t \geq 10 \quad \text{or} \quad D_0/t < 10$$

Where

D_0 = External diameter of the vessel

t = Minimum required thickness of the cylindrical shell

L = Design length of a vessel section determined per Code Paragraph UG-28(b)

Thickness is first determined from the internal pressure; then the external pressure calculation is used to determine the following:

If the thickness is adequate for the external design pressure, or

If additional stiffening of the shell is required

If the thickness of the shell has to be determined to start the procedure, a value for t must be assumed

The code procedure goes through several steps for each alternative in order to find the value of Factor "B," determined by using a general geometric chart in Subpart 3, Part D, of Section II

For $D_0/t \geq 10$, the maximum allowable external pressure is determined using the following equation:

$$P_a = \frac{4B}{3(D_0/t)} \tag{6}$$

For $D_0/t < 10$, two values, P_{a1} and P_{a2}, should be determined, and the lower value is used:

$$P_{a1} = \left[\frac{2.167}{(D_0/t)} - 0.0833\right]B \tag{7}$$

and

$$P_{a2} = \frac{2S}{(D_0/t)}\left[1 - \frac{1}{(D_0/t)}\right] \tag{8}$$

Where S is the lesser of

Two times the allowable stress in tension or

0.9 times the yield strength of the material

Example

Use the same vessel as in the internal pressure calculation.

Given

Tangent to tangent length = 36 ft = 432 inches

Two 2:1 semi-ellipsoidal heads

External design pressure: 15 psig @ 500°F

L = 448 inches (length of shell plus one third of the depth of each head, 16 inches)

$L/D_0 = 448/97 = 4.62$

$D_0t/ = 97/0.375$ (corroded) = 258.67

Determine

Evaluate the vessel for external pressure.

Solution

1. Enter Figure G, Section II, Part D, Subpart 3 with $L/D_0 =$ 4.62 and read across to sloping line of $D_0/t = 258.7$. Read factor A = 0.00007.
2. Enter Figure CS-2, Section II, Part D, Subpart 3 with A = 0.00007, which is off to the left side and cannot be read; modulus of elasticity of material, $E = 27 \times 10^6$ psi
3. For values falling to the left of the material/temperature line, P_a can be determined from the following:

$$P_a = \frac{2AE}{3(D_0/t)} = \frac{(2)(0.00007)(27 \times 10^6)}{3(258.67)} = 4.87\ psi$$

The vessel is good for only 4.87 psi external pressure. To increase the vessel's resistance to external pressure, we must either increase vessel thickness or require stiffening rings.

4. Determine if stiffening ring spacing is done by trial and error. Try two stiffening rings equally spaced between tangent lines:

$$\begin{aligned} L &= 144 \text{ (length of shell between rings)} \\ &\quad +8 \text{ (one third the depth of head)} \\ &= 152 \text{ inches} \end{aligned}$$

$$L/D_0 = 152/97 = 1.56$$

$$A = 0.00022 \text{ from Figure 4.6 with}$$

$$L/D_0 = 1.56 \text{ and } D_0/t = 258.7$$

$$B = 2800 \text{ from Figure CS} - 2$$

Using the value of B, we calculate the value of the maximum allowable external pressure, P_a, from the following equation:

$$P_a = \frac{4B}{3(D_O/t)} = \frac{(4)(2800)}{3(258.67)} = 14.4\,psi$$

The vessel is still not good for 15 psig, so another ring should be added.

Solution

5. Stiffening ring spacing determination:

   Try three stiffening rings equally spaced between tangent lines:

$$L = 108 + 8 = 116 \text{ inches}$$

$$L/D_O = 116/97 = 1.19$$

$$A = 0.00027 \text{ from Figure G with}$$

$$L/D_O = 1.19 \text{ and } D_O/t = 258.67$$

$$B = 3700 \text{ from Figure CS} - 2$$

Using the value of B, we calculate the value of the maximum allowable external pressure, P_a, from the following equation:

$$P_a = \frac{4B}{3(D_O/t)} = \frac{(4)(3700)}{3(258.67)} = 19.07\,psi$$

Because P_a is greater than the design pressure, the vessel with three stiffening rings is good for full vacuum service (15 psi).

Design of Stiffening Rings

The equations for the maximum strength of a cylindrical shell under external pressure are established under the assumption that the shell is simply supported

For this to be true, stiffening rings are used as lines of support, assumed to carry the entire load that the shell carries because of external pressure

The size of the stiffening ring is calculated by using the equation for buckling of a cylinder ring under external pressure

The required moment of inertia of the stiffening ring is shown by the following equation:

$$I = \frac{D_0^2/L_s(t + A_s/L_s)A}{14} \tag{9}$$

Where:

L_s = The sum of one half the distances on both sides of the stiffening ring from the center line of the ring (1) to the next stiffening, (2) to a circumferential line on the head line at one third of its depth, or (3) to a jacket connection

A_s = Cross-sectional area of the stiffening ring

Code Paragraph UG-29 allows a portion of the shell to be considered as contributing to the moment of inertia

The width of the shell section is 1.1 D_0 t and is considered as laying one half on each side of the centroid of the ring

In this case, the required moment of inertia of the combined area is as follows:

$$I = \frac{D_0^2/L_s(t + A_s/L_s)A}{10.9} \tag{10}$$

The code procedure for determining the size of the ring is as follows:

1. Assume a ring size; determine A_s
2. Calculate Factor B

$$B = \left(\frac{3}{4}\right)\frac{PD_0}{t + A_s/L_s} \tag{11}$$

3. From Chart G in Subpart 3, find the value of A
4. Calculate the value of I_s; the moment of inertia of the ring or the combined section (ring plus shell section) has to be larger than the required moment of inertia

Example

Using the same vessel as in the previous example, select a ring size and determine if the vessel is adequately stiffened.

Solution

1. Selection a 5 × 3 × 3/8 angle ring with $A_s = 2.86$ inches^2 and $I = 7.37$ inches^4.
2. Calculate Factor B.

$$B = \left(\frac{3}{4}\right)\frac{PD_0}{t + A_s/L_s} = \left(\frac{3}{4}\right)\frac{(15)(97)}{(0.375) + (2.68/116)}$$
$$= 2730 \qquad (12)$$

3. From Chart G in Section 3, A = 0.0002 with B = 2730
4. Calculate I_s.

Because the available moment of inertia is larger than required, the vessel is adequately stiffened

Spherical Shell and Hemispherical Heads under External Pressure

Example

The ASME code procedure is as follows:

1. Calculate the value of A using the equation

$$A = \frac{0.125}{R_0/t} \qquad (13)$$

Where:

R_0 = Outside radius of the sphere

2. Find the value of B from Figure CS-2 in Subpart 3.

3. Calculate the maximum allowable external pressure by using the following equation:

$$P_a = \frac{B}{R_0/t} \tag{14}$$

Or for values of A falling to the left of applicable temperature line:

$$P_a = 0.0625E/(R_0/t)^2 \tag{15}$$

Solution

4. Calculate I_s.

$$I_s = \frac{D_0^2 L_s(t + A_s/L_s)A}{14}$$

$$= \frac{(97)^2(116)(0.375 + 2.86/116)(0.0002)}{14} = 6.23\ in^4$$

Because the available moment of inertia is larger than required, the vessel is adequately stiffened.

Example

What external design pressure is a hemispherical head good for using the following design conditions?

Solution

1. For $R_0 = 48.5$ inches and t = 0.125 inch,

$$= \frac{0.125}{48.5/0.125} = 0.0013$$

2. From Figure CS-2 in Subpart 3, B = 10,500 with A = 0.0013.
3. Calculate the maximum allowable external pressure:

$$P_a = \frac{B}{R_0/t} = \frac{10{,}500}{48.5/0.187} = 54.13$$

The hemispherical head is good for the external design pressure of 54 psi.

Semi-ellipsoidal Heads under External Pressure

According to the code, the required thickness for a semi-ellipsoidal head under external pressure should be the greater of the following:

1. The thickness as calculated by the equation given for internal pressure using a design pressure 1.67 times the external pressure and joint efficiency, E = 1.00
2. The thickness by the equation $P_a = B/(R/t)$, where $R = 0.9D_0$ and B is to be determined for a sphere

Torispherical Heads under External Pressure

The required thickness is computed by the procedure given for ellipsoidal heads using a value for $R = D_0$

Example of Internal/External Pressure Design

Problem Statement

Determine the minimum required thickness of a cylindrical shell and hemispherical heads of a welded pressure vessel designed for an internal pressure of 100 psi at a design temperature of 250°F. There is no corrosion. The shell, which contains a longitudinal butt weld, is also butt welded to seamless heads. All Category A butt joints are type (1) with full radiography (RT). E = 1.00 for all calculations. The shell has a 5-foot, 0-inch inside radius and a 30-foot, 0-inch length from tangent to tangent.

Also determine the minimum required thickness of the same vessel designed for an external pressure of 15 psi at 100°F without stiffening rings. What is the stiffening ring spacing if the required thickness of internal pressure is used?

Solution

1. For SA 515 Grade 50, the allowable tensile stress from Table 1A of Section II, Part D at 100°F is 15.0 ksi, and the external pressure chart is Figure CS-2 of Subpart 3 of Section II, Part D.
2. As is generally the case for internal pressure on a cylinder, when E = 1.00 for all butt joints, UG-27(c)(1) for circumferential stress (hoop stress) controls over UG-27(c)(2) longitudinal stress by the following:

$$t = \frac{PR}{SE - 0.6\,P} = \frac{(100)(60)}{(15,000)(1.0) - (0.6)(100)}$$

$$= 0.401\ in$$

Check for applicability of using UG-27(c)(1):
Is t < R/2; 0.400 < 30 o.k.
Is P < 0.386 SE ; 100 < 0.385(15,000)(1) = 5775 o.k.

3. For internal pressure on hemispherical heads, use UG-32(f):

$$t = \frac{PL}{2SE - 0.2\,P} = \frac{(100)(60)}{2(15,000)(1.0) - (0.2)(100)} = 0.200\ in$$

Is t < 0.365 *L*; 0.200 < .365(60) = 21.9 o.k.

Is P < 0.665 SE : 100 < 0.665(15,000) (1) = 9975 o.k.

4. For external pressure on a cylinder, use UG-28 and Subpart 3.

For cylindrical shells with formed heads on the end, the length of the shell plus one third of the depth of each head is used to determine the effective lengths (L) (see UG-28).

Determine the effective length without stiffening rings = one third of each head depth plus straight length = 2(1/3)(60) + 360 = 400 inches.

Assume t_{min} = 0.400 for internal pressure and D_0= 120 + 2(.4).

Then

$$L/D_0 = 400/120.8 = 3.31$$

$$D_0/t = 120.8/0.4 = 301$$

a. Enter Figure G with $L/D_0 = 3.31$ and read across to sloping line of $D_0/t = 301$. Read Factor A = 0.00075.

b. Enter Figure CS-2 with A = 0.00075 and the modulus of elasticity $E = 29 \times 10^6$, which is off the left side and cannot be read.

Following step (7) of UG-28(c):

$$P_a = \frac{2AE}{3(D_0/t)} = \frac{2(0.000075)(29 \times 10^6)}{3(301)}$$

$$= 4.817 \text{ psi} < 15 \text{ psi therefore must increase thickness}$$

Assume t = 5/8 inch = 0.625 inch and D_0 = 120 + 2(0.625) = 121.25 inches.

Then

$$L/D_0 = 400/121.25 = 3.30$$

$$D_0/t = 121.25/0.625 = 194$$

c. From Figure G, Factor A = 0.00014.

d. Recalculating P_a:

$$P_a = \frac{2(0.00014)(29 \times 10^6)}{3(194)}$$

$$= 14 \text{ psi} < 15 \text{ psi} \therefore \text{must increase thickness}$$

Assume t = 11/16 inch = 0.6875 inch and $D_0 = 120 + 2(0.6875) = 121.375$ inches.

Then

$$L/D_0 = 400/121.375 = 3.30$$

$$D_0/t = 121.375/0.6875 = 177$$

e. From Figure G, Factor A = 0.00017.

f. Recalculating P_a:

$$P_a = \frac{2(0.00017)(29 \times 10^6)}{3(177)}$$

$$= 18.6 \text{ psi} > 15 \text{ psi external pressure}$$

Further calculations show that $t_{min} = 0.64$ inch for 15 psi external pressure.

5. For external pressure on the hemispherical head, use UG-33(c), UG-28(d), and Subpart 3:

   First assumption, use t_{min} for internal pressure of t = 0.200.

   Assume t = 0.200 inch and $R_0 = 0.5(120 + 2)0.2 = 60.2$ inches

   a. Calculate A:

$$A = \frac{0.125}{R_0/t} = \frac{0.125}{(60.2/0.2)} = 0.0004$$

b. Enter Figure CS-2 with Factor A = 0.0004 and read B = 5800.

c. Determine P_a:

$$P_a = \frac{B}{(R_0/t)} = \frac{5800}{(60.2/0.2)}$$
$$= 19.4 \text{ psi} > 15 \text{ psi MAWP}$$

Further calculations show that t_{min} = 0.17 inch for 15 psi.

Of interest is the fact that for 100 psi internal pressure the minimum required thickness of the cylinder is 0.401, whereas for 15 psi external pressure the minimum required thickness is 0.636 inch. For the head, the minimum required thickness is only 0.200 inch. for internal pressure, whereas for external pressure the minimum required thickness is less than 0.200 inch.

If a thickness between 0.400 and 0.636 inch is desired for the cylinder, stiffening rings are required on the cylinder to obtain a smaller value of L to use in the calculation of P_a. By trial and error, the approximate maximum stiffening ring spacing with the minimum thickness required for internal pressure of 0.400 is 120 inches as follows:

Assume t = 0.400 inch and L = 120 inches

Then

$$L/D_0 = 120/120.8 = 0.993$$

$$D_0/t = 302$$

d. Enter Figure G, read Factor A = 0.00025.

f. Enter Figure CS-2, read Factor B = 3500.

g. Determine P_a:

$$Pa = \frac{4B}{3(D_O/t)} = \frac{(4)(3500)}{3(302)} = 15.45 \text{ psi}$$

$$> 15.0 \text{ psi MAWP}$$

This indicates that the optimum design would be one where the shell was thickened above 0.400 inch with stiffening rings being placed at a spacing larger than 120 inches center to center. The optimum design would be obtained by trial and error. After the “best” thickness and stiffening ring spacing is determined, the design of the stiffening ring is developed according to UG-29.

Openings and Nozzle Reinforcement

Vessel Openings

Openings are required to do the following:
- Attach piping, mechanical equipment, and instrumentation
- Permit inspections

Physical boundary between the jurisdiction of the ASME Code and the appropriate piping code is one of the following interfaces:
- Welded pipe: between the first circumferential joint of pipe and the nozzle (vessel shipped without flange)
- Screwed connections: first threaded joint
- Flanged connections: first flange face (shipped with flange)
- Other connections: first sealing surface

Opening Shapes and Sizes

Code permits a pressure vessel to be designed with openings of any shape or size

Circular and elliptical openings are recommended because vessel manufacturers are familiar with these types

An "obround" opening is formed by two parallel sides and semicircular ends; provides good access for maintenance personnel

Large head room

Code does not restrict the size of a vessel opening; large openings require reinforcement, which may cause cost and fabrication problems

Good design practice dictates that when an opening becomes more than half the inside diameter of the shell, the design should employ a self-reducing section instead of a nozzle

Inspection Openings

Required by the code for all pressure vessels containing process environments that cause corrosion, erosion, or mechanical abrasion

Vessels that contain internals usually require maintenance of the internals during the life of the unit

Should be designed to permit reasonable entry for personnel, welding equipment, and internal components

Code requires a 15-inch manhole in vessels having an inside diameter of more than 56 inches

The following guidelines should be observed in most cases:

1. For vessels 12-inch nominal diameter and smaller, a means of inspection is often omitted
2. Vessels from 12-inch to 18-inch nominal diameter should have two inspection openings of 2½-inch minimum diameter (the code only requires two 2-inch openings)
3. Vessels from 18-inch through 36-inch nominal diameter should have two flanged 4-inch openings or a manhole (the code only requires two 2-inch openings)

4. If replacement of internals is necessary, one end of the vessel should have a full diameter flange for vessel sizes through 24-inch diameter; from 24-inch through 36-inch vessel diameter, one of the following can be considered:
   A conical section and a 20-inch diameter flanged end (generally the least expensive and most satisfactory alternative)
   Full diameter flanged end
   A shell or head manhole
5. Vessels larger than 36 inches in diameter
   All should have at least one manhole; recommended minimum diameter is 18 inches, although the code allows a 15-inch diameter
   In relatively large vessels, two manholes are frequently provided to facilitate maintenance and to improve ventilation
6. The following practices assume that access through internal tray manholes is not restricted by appurtenances or by staggering of the tray manholes:
   Columns with 10 trays or fewer should be equipped with a manhole below the bottom tray and above the top tray
   Columns with more than 10 trays but fewer than 30 trays should generally be equipped with a manhole below the bottom tray, above the top tray, and near the middle of the column, preferably near the feed tray
   Columns of more than 30 trays present access problems that must be given special review; manholes should be provided below the bottom tray and above the top tray; intermediate manholes should, in general, be placed no farther than 30 feet (about 15 trays)

apart and located as close as economical to points of high corrosion, such as feed trays; if possible, project or plant design specifications should establish manhole locations for each column

The relation between limitations on maximum ladder height without a break (30 feet) and the normal limit for number of trays between manholes (15) should be reviewed for balance between safety and economy

7. Normally, manholes should be at least 18-inch nominal diameter; this diameter is sufficient to accommodate routine inspection and equipment for minor maintenance; however, if internal materials, equipment, and so on require extensive maintenance or internal staging, a 24-inch manhole should be provided to facilitate access of personnel and equipment

Nozzle Neck Thickness

The wall thickness of a nozzle neck or other connection should not be less than the thickness computed for the applicable loadings plus the thickness added for corrosion allowance on the connection

Except for access openings and openings for inspection only, it should not be less than the smaller of the following:

1. The minimum thickness of standard wall pipe plus corrosion allowance on the connection
2. The required thickness (assuming $E = 1)^{+}$ of the shell (or head) to which the connection is attached plus the corrosion allowance provided in the shell (or head) adjacent to the connection; but for a welded vessel, in no case less than 1.16 inches

Note: E = 0.80 when the opening in a vessel is not radiographed

The minimum thickness of a pipe is the nominal wall thickness less the 12½% allowable mill tolerance

Concept of Reinforcing Openings

When an opening is made in a pressure vessel shell or head, the ability of the nearby wall to retain pressure is significantly reduced

Reinforcing pressure vessel openings maintains the pressure retaining capabilities of the shell by the addition of wall thickness near the opening

The basic rule of the code is that the wall section around the opening of the vessel must be reinforced with an area of metal equal to the area of metal removed to create the opening

The replaced area of material is called the opening or nozzle reinforcement

The reinforcement may be incorporated into the following:

- Vessel wall
- Nozzle wall
- Attached pad surrounding the nozzle

This simple code rule needs amplification:

1. The code says it is not necessary to replace the removed amount of metal, but the amount of wall thickness required to resist the internal pressure; this required thickness at the openings is usually less than at other points of the shell or head, because of corrosion allowance and nominal size plates yielding extra thickness; in computing the allowable pressure for an existing vessel, most engineers follow the practice of using the lesser wall thickness at the openings, but when designing new vessels, openings can often be reinforced to full as-built

shell thickness so vessels may be used up to the limits of their strength whenever required; in the design of new vessels, there will be no appreciable extra cost if the openings are reinforced for the full thickness of the new plate

2. The plate actually used and nozzle neck are normally thicker than would be required according to calculation; according to the code, the excess plates in the vessel wall (A_1) and nozzle wall (A_2) serve as reinforcements (refer to Figure 4.6)

   Likewise, the inside extension of the opening (A_3) and the area of the weld metal (A_4) can also be taken as credit for reinforcement; the recommended practice is to assume that A is equal to zero—that is, no excess in vessel wall is credited to reinforcement; instead, the excess should be assigned to corrosion allowance
3. The reinforcement must be within certain, dimensional limits
4. The area of reinforcement must be proportionally increased if its stress value is lower than that of the vessel wall
5. The area requirement for reinforcement must be satisfied for all planes through the center of opening and normal to the vessel surface

Figure 4.6 presents the equations and terminology for determining nozzle reinforcement

Calculations for Reinforcement of Openings

To determine whether an opening is adequately reinforced, it is first necessary to determine whether the areas of reinforcement available will be sufficient without the use of a pad

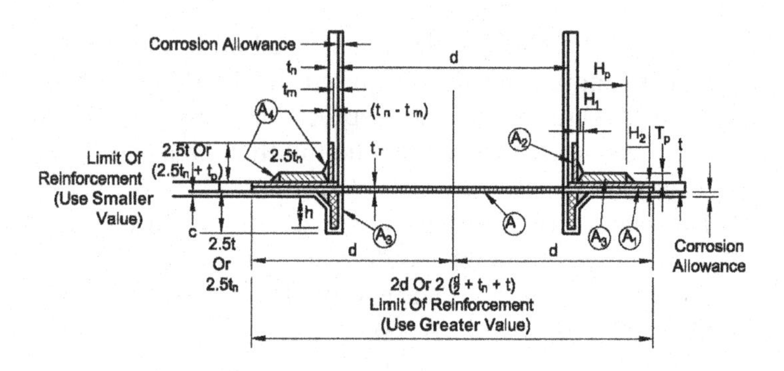

NOZZLE DATA2

P = Pressure, psi

S = Allowable stress, psi

R_n = Inside radius of nozzle, in

E_1 = Joint efficiency, %

$$t_m = \frac{PR_n}{SE_1 - 0.6P}$$

t_n = Actual thickness of nozzle (after subtracting the corrosion allowance)

t_m = Calculated required thickness of nozzle

$t_n - t_m$ = Excess thickness in nozzle

SHELL DATA

t = Actual thickness of shell or head (minus corrosion allowance)

tr = Calculated thickness of shell or head

$E_1t - Ft_r$ = Excess thickness in shell or head

Area of reinforcement required $A = dt_rF$

Area of excess thickness in shell or head (use greater value) $A_1 = (E_1t - Ft_r)d$, or

Figure 4.6 Nozzle reinforcement calculation.

The total cross-sectional area of reinforcement required (inches^2) is indicated by the letter A, which is equal to the diameter (plus corrosion allowance) times the required thickness

(Chevron practice is to set $A_1 = 0$)	$A_1 = 2\,(E_1t - Ft_r)\,(t + t_n)$
Area available in nozzle projecting inward	$A_3 = 2\,(t_n\text{-}c)\,f_r \times h$
Cross-sectional area of weld	$A_4 = 2\left[\frac{(W_1)^2 + (W_2)^2}{2}\right] \times F_r$
Area of reinforcement available without pad	$(A_1 + A_2 + A_3 + A_4)$
If $A_1 + A_2 + A_3 + A_4 \geq A$	Opening is adequately reinforced.
If $A_1 + A_2 + A_3 + A_4 < A$	Opening is not adequately reinforced.

Element must be added or thickness Increased.

Area is pad $A_\sigma = 2W_pT_p$ = ____________________

Total area available = ____________________

1. if reinforcing pad is used, the factor $2.5t_n$ becomes $(2.5t_n + Tp)$

2. For cases when the allowable stress of the nozzle of reinforcing element is less than allowable stress of the vessel, refer to Code par. UG-41 (A) and appendix L, latest Addenda, for consideration of this effect

Figure 4.6 *(continued)*.

The area of reinforcement available without a pad includes the following:

The area of excess thickness in the shell or head, A_1 (usual practice is to set $A_1 = 0$ and allocate it to corrosion allowance)

The area of excess thickness in the nozzle wall, A_2

The area available in the nozzle projecting inward, A_3

The cross-sectional area of welds, A_4

If $A_1 + A_2 + A_3 + A_4 \geq A$, the opening is adequately reinforced

If $A_1 + A_2 + A_3 + A_4 \leq A$, a pad is needed

If the reinforcement is found to be inadequate, then the area of pad needed (A_5) may be calculated as follows:

$$A_5 = A - (A_1 + A_2 + A_3 + A_4) \quad (16)$$

Refer to Figure 4.6 for nozzle and shell data, equations, and nomenclature for calculating the area of pad needed for reinforcement

If a pad is used, the factor (2.5t) in the equation for A_2 in Figure 4.6 is measured from the top surface of the pad and therefore becomes ($2.5t_n + T_p$)

The area A_2 must be recalculated on the basis and the smaller value again used; then

If $A_1 + A_2 + A_3 + A_4 + A_5 \geq A$, the opening is adequately reinforced

The other symbols in Figure 4.6 for the area equations (all values except E_1 and F are in inches) are as follows:

- d = Diameter in the plane under consideration of the finished opening in its corroded condition
- t = Nominal thickness of shell or head, less corrosion allowance
- t_r = Required thickness of shell or head as defined in Code Paragraph UG-37
- t_{rn} = Required thickness of a seamless nozzle wall
- T_p = Thickness of reinforcement pad
- W_p = Width of reinforcement pad
- t_n = Nominal thickness of nozzle wall, less corrosion allowance
- W_1 = Cross-sectional area of weld
- W_2 = Cross-sectional area of weld
- E_1 = 1, when an opening is in the plate or when the opening passes through a circumferential point in a shell or cone (exclusive of head-to-shell joints), or

Joint efficiency obtained from Code Table UW-12 when any part of the opening passes through any other welded joint

F = Correction factor that compensates for the variation in pressure stresses on different planes with respect to the axis of vessel; a value of 1 is used for F in all cases except when the opening is integrally reinforced; if integrally reinforced, see Code Figure UG-37*

f_r = Strength reduction factor; ratio of material stressor: ≤ 1

h = Distance nozzle projects beyond the inner surface of the vessel wall before corrosion allowance is added

To correct for corrosion, deduct the specified allowance from shell thickness, t, and nozzle thickness, tn, but add twice its value to the diameter of opening d

On all welded vessels built under Column C of Code Table UW-12, 80% of the allowable stress value must be used in design equations and calculations

Figures 4.7, 4.8, and 4.9 provide examples of reinforcement calculations

Design for External Pressure on Openings

The reinforcement required for openings in single-walled vessels subject to external pressure need be only 50% of that required for internal pressure, where tr is the wall thickness required by the rules for vessels under external pressure (Code UG-37(c)(1))

* Reinforcement is considered integral when it is inherent in the shell plate or nozzle. Reinforcement built up by welding is also considered integral. Installation of a reinforcement pad is not considered integral.

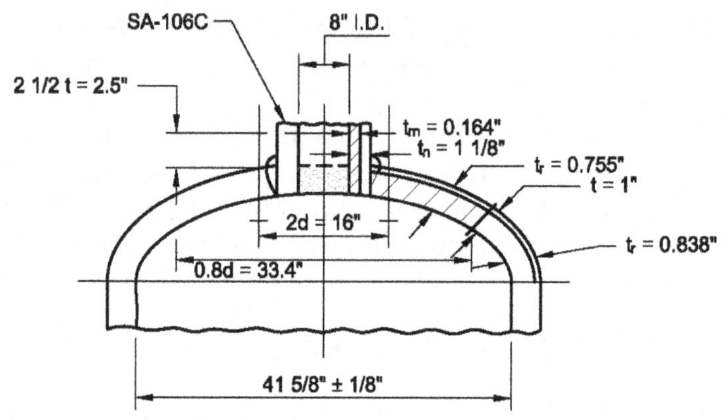

Figure 4.7 Example 1: reinforcement calculations.

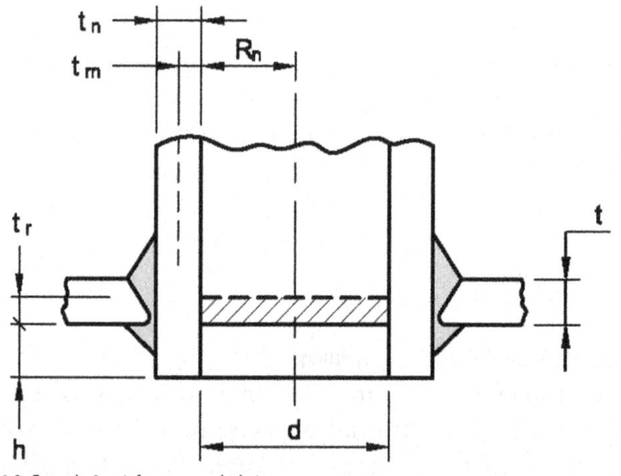

Figure 4.8 Example 2: reinforcement calculations.

Example 1 Reinforcement Calculations

Problem

Determine the reinforcement requirements of an 8-inch I.D. nozzle that is centrally located in a 2:1 ellipsoidal head.

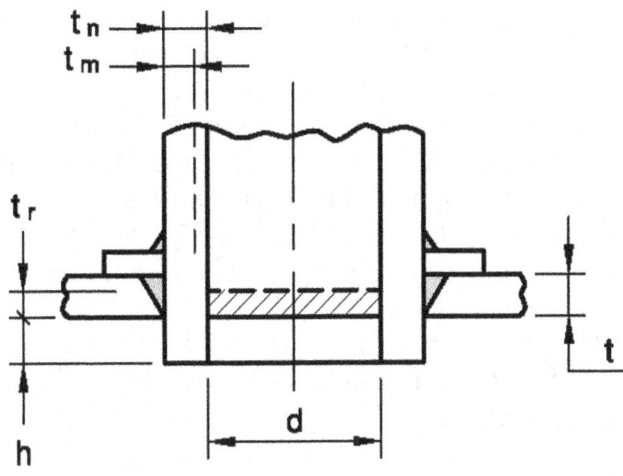

Figure 4.9 Example 3: reinforcement calculations.

The inside diameter of the head skirt is 41.75 inches. The head material is SA 516 Gr 70, and the nozzle is SA 106 Gr C. The design pressure is 700 psig. The design temperature is 500°F. There is no corrosion, and the weld joint efficiency is 1.

Solution

1. Because the allowable tensile stress for both SA 516 Gr. 70 and SA 106 Gr. C at 500°F is 17.5 ksi, the material strength reduction factor is $f_r = 1$.
2. The minimum required thickness of a 2:1 ellipsoidal head without an opening is determined from UG-32(d) as follows:

$$t = \frac{PD}{2SE - 0.2P} = \frac{(700)(41.75)}{2(17,500 \times 1.0) - 0.2(700)}$$

$$= 0.838\ in.\ \text{Actual thickness used is 1 inch.}$$

3. According to Rule (3) for t_r in UG-37(a), when an opening and its reinforcement are in an ellipsoidal head and are located entirely within a circle the center of which coincides with the head and the diameter is equal to 80% of the shell diameter, t_r is the thickness required for a seamless sphere of radius K_1D, where D is the shell I.D. and K_1 is 0.9 from Table UG-37. For this head, the opening and reinforcement shall be within a circle with a diameter of 0.8d = (0.8)(41.75) = 33.4 inches.
4. Following (3) above, the radius $R = K_1D = 0.9(41.75) = 37.575$ inches is used in UG-27(d) to determine the t_r for reinforcement as follows:

$$t_r = \frac{PR}{2SE - 0.2P} = \frac{(700)(37.575)}{2(17{,}500 \times 1.0) - 0.2(700)}$$
$$= 0.755\ in.$$

5. Using UG-27(c)(1), determine the required nozzle thickness:

$$t_{rn} = \frac{PR_n}{2SE - 0.2P} = \frac{(700)(4)}{(17{,}500 \times 1.0) - 0.6(700)}$$
$$= 0.164\ in.$$

6. Limit parallel to head surface = X = d or ($d/2 + t + t_n$), whichever is larger; X = 8 inches or (4 + 1 +1.125 = 6.125), use X = 8 inches.
7. Limit perpendicular to shell surface = Y = 2½t or 2 ½t_n. whichever is smaller. Y = 2 1/2(1) = 2.5 inches or 2½(1.125) = 2.81 inches, use 2.5 inches.
8. Limits of 2X = 2(8) = 16 inches is less than 33.4 inches; therefore, the provision to use spherical head rule is valid.

9. Reinforcement area required according to UG-37(c) is as follows:

$$A_1 = dt_r F + 2t_n t_r F(1 - f_{r1}) = (8)(0.755)(1) + 0$$
$$= 6.040 \text{ inches}^2 \text{ when } f_{r1} = 1$$

10. Reinforcement available in head is as follows:

$$A_1 = d(Et - Ft_r) - 2t_n(Et - Ft_r)(1 - f_{r1})$$

When $f_{r1} = 1$, the second term becomes zero; therefore, for $E = 1$ and $F = 1$:

$$A_1 = d(t - t_r) = (8)(1.0 - 0.755) = 1.960 \text{ inches}^2$$

11. Reinforcement available in the nozzle is as follows:

$$A_2 = 2Y(t_n - t_{rn}) = (2)(2.5)(1.125 - 0.164)$$
$$= 4.805 \text{ inches}^2$$

12. Total reinforcement available in head and nozzle is as follows:

$$A_t = A_1 + A_2 = 1.960 + 4.805 = 6.765 \text{ inches}^2$$

Area available of 6.765 inches^2 is greater than area required of 6.040 inches^2.

13. Determination of weld strength and load paths. According to UW-15(b), strength calculations for welds for pressure loading are not required for nozzles like that shown in Figure UW-16.1(c). Because this nozzle is similar to that detail, no load path calculations are required.

Example 2 Reinforcement Calculations

Design Data

Inside diameter of shell, D = 48 inches
Design pressure, P = 300 psi at 200°F
Shell material, t = 0.500 inches SA 516-70 plate
S = 17,500 psi
The vessel spot examined, E = 0.85
There is no allowance for corrosion.
Nozzle nominal size 6 inches $R_n = 2.88$
Nozzle material, $t_n = 0.432$-inch wall, SA 53B, seamless pipe
S = 15,000 psi
Extension of nozzle inside the vessel, 1.5 inches $h = 2.5t_n = 2.5 \times 0.432 = 1.08$ inches
The nozzle does not pass through the main seams.
Fillet weld size 0.500 inch
Wall thickness required:

$$Shell\ t_r = \frac{PR}{SE - 0.6P} = \frac{300 \times 24}{17{,}500 \times 1 - 0.6 \times 300}$$

$$= 0.416 \text{ inch}$$

$$Nozzle\ t_{r_n} = \frac{PR_n}{SE - 0.6P}$$

$$= \frac{300 \times 2.88}{15{,}000 \times 1 - 0.6 \times 300}$$

$$= 0.058 \text{ inch}$$

Area of reinforcement required:

$$A = dt_r = 5.761 \times 0.416 = 2.397 \text{ inches}^2$$

Area of reinforcement available: *A_1 = (excess in shell) larger of following:

$(t - tr)d = (0.500 - 0.416)\ 5.76$ inches^2

or $(t - t_r)(t_n + t)2 = (0.500 - 0.416)\ (0.432 + 0.500)\ 2 =$ 0.156 inch^2

A_2 = (excess in nozzle neck) smaller of following:

$(t_r - t_{rn})\ 5t = (0.432 - 0.058)\ 5 \times 0.500 = 0.935$

or $(t_n - t_{rn})\ 5t_n = (0.432 - 0.058)\ 5 \times 0.432 = 0.808$

Because the strength of the nozzle is lower than that of the shell, a decreased area shall be taken into consideration:

$$15{,}000/17{,}500 = 0.857, 0.857 \times 0.808 = 0.692 \text{ inch}^2$$

$$A_3 = \text{(inside projection) } t_n \times 2h = 0.432 \times 2 \times 1.08 = 0.933$$

$$\text{Area decreased } 0.933 \times 0.857 = 0.800 \text{ inch}^2$$

$$A_4 = \text{(area of fillet weld) } 2 \times 0.5 \times 0.5002 \times 0.857 = 0.214 \text{ inch}^2$$

$$A_5 = \text{(area of fillet weld inside) } 2 \times 0.5 \times 0.5002 \times 0.857 = 0.214 \text{ inch}^2$$

Total area available:

Because this available area is greater than the area required for reinforcement, additional reinforcement is not needed.

*The recommended company practice would assign a value of 0 to A_1. This example illustrates minimum code requirements.

Example 3 Reinforcement Calculations

Problem

Determine the reinforcement requirements for an 8-inch SA-53B seamless schedule 80 radial 0.500-inch thick seamless pipe nozzle projecting 1.25 inches into the vessel I.D., in a 0.500-inch thick cylindrical pressure vessel shell of inside radius of 24 inches with a design pressure of 300 psig at 200°F. The vessel is fully radiographed and E = 1.0 for all joint efficiencies. There is no corrosion and the nozzle does not intersect any main seams. It has 0.375-inch fillet welds.

1. The required solution wall thicknesses are as follows:

$$Shell\ t_r = \frac{PR}{SE - 0.6P} = \frac{300 \times 24}{17{,}500 \times 1 - 0.6 \times 300}$$

$$= 0.416 \text{ inch}$$

$$Nozzle\ t_{rn} = \frac{PR_n}{SE - 0.6P} = \frac{300 \times 3.8125}{15{,}000 \times 1 - 0.6 \times 300}$$

$$= 0.077 \text{ inch}$$

2. Area of reinforcement required:

$$A = d \times t_r = 7.625 \times 0.416 = 3.172 \text{ inches}^2$$

3. Limits of reinforcement:

Parallel to shell

$$X = d \text{ or } r_n + t_n + t, \text{ the larger}$$

$$= 7.625 \text{ or } 3.8125 + 0.5 + 0.5 \qquad \text{Use } X = 7.625$$

Perpendicular to shell

$$Y = 2^1\!/\!_2\ t \text{ or } 2^1\!/\!_2 t_n, \text{ the smaller,}$$

since both have a t of 0.500 inch

$$= 1.25 \text{ or } 1.25 \quad \text{Use } Y = 1.25$$

4. Area of reinforcement available:

*A1 = (excess in shell) larger of following:

$$(t - t_r)d = (0.500 - 0.416)7.625 = 0.641\ \text{inch}^2$$

$$\text{or } (t - t_r)(t_n + 1)2 = (0.500 - 0.416)(0.500 + 0.500)\ 2$$
$$= 0.168^2$$

A^2 = (excess in nozzle neck) Smaller of following:

$$(t_n - t_{rn})\ 5t_n = (0.500 - 0.077)\ 50.5 = 1.058^2$$

$$\text{or } (t_n - t_{rn})\ 5t_n = (0.500 - 0.077)\ 5 \times 0.5 = 1.058^2$$

Because the strength of the nozzle is lower than that of the shell, a reduced area shall be taken into consideration:

$$15{,}000/17{,}500 = 0.857,\ 0.857 \times 1.058^2 = 0.907\ \text{inch}^2$$

$$A_3 = \text{(inside projection) } t_n \times 2h = 0.5 \times 2$$

$$\text{Area decreased } 0.857 \times 1.25 = 1.071\ \text{inch}^2$$

$$A_4 = \text{(area of fillet weld) } 2 \times 0.5 \times 0.375^2 = 0.141\ \text{inch}^2$$

(The area of pad-to-shell weld is disregarded)

Total area available: = 2.76 inches^2

This area is less than the required area; therefore, the difference shall be provided by a reinforcing element. It may be heavier nozzle neck, larger extension of the nozzle inside of the vessel, or reinforcing pad. Using a reinforcing pad, the required area of pad is $3.172 - 2.760 = 0.412\ \text{inch}^2$.

Using 0.250 in SA 516-70 plate for re-pad the width of the pad:

$$\frac{0.412}{0.25} = 1.65$$

The outside diameter of reinforcing pad:

Outside diameter of pipe:	8.625
Width of reinforcing pad:	1.648
	10.273 in.

Reinforcement of Openings for External Pressure

The cross-sectional area (A) of reinforcement required for openings in vessels subject to external pressure is as follows:

$$A = \frac{dt_rF}{2} \tag{17}$$

Where

d = Diameter in the given plane of the opening in its corroded condition, inches

t_r = Wall thickness required for external pressure, inches

F = Factor for computation of the required reinforcement area on different planes (as the pressure-stress varies) when the opening is in a cylindrical shell or cone and integrally reinforced; for all other configurations, the value of F = 1

Bolted Flanged Connections

General Considerations

The code covers the design of flanges in (mandatory) Appendix 2

The scope of the rules applies to gaskets contained entirely within the bolt circle

According to the rules, acceptable flanged nozzles may be attained by the use of either of the following:
Standard rated flanges

Flange calculations

Normally, the standard ANSI B16.5 "Pipe Flanges" or API 605 "Large Diameter Carbon Steel Flanges" are used as they can be selected with very little design effort—that is, "off-the-shelf"

When a standard flange is selected from these specifications, no additional calculations are required

The following are typical flange standards:

- MSS SP-44, Classes 300, 400, 600, and 900 in sizes 26 to 36 inches
- Standard 605, "Large Diameter Carbon Steel Flanges," 75-, 150-, and 300-pound rating in sizes 26 to 60 inches inclusive
- Taylor Forge Standard, Classes 75, 175, and 350 in sizes 26 to 72 inches, 92 and 96 inches
- American Water Works Association (AWWA) Standard C207-55, Classes, B, D, and E, in sizes 6 to 96 inches

Follow the rules of Code Appendix 2 for the following conditions:

1. When it is necessary to calculate a flange because one of the standard flanges in the correct size is not available
2. When the pressure temperature ratings are not adequate
3. When special design considerations are to be addressed

The calculation of flanges requires the following:

- Selection of materials for flange, bolts, and gaskets
- Determination of facing and gasket details so that the bolt loading may be determined and bolt sizes selected
- Determination of the bolt circle and the loads, moment arms, and moments because of gasket setting and operating conditions

Determination of stresses

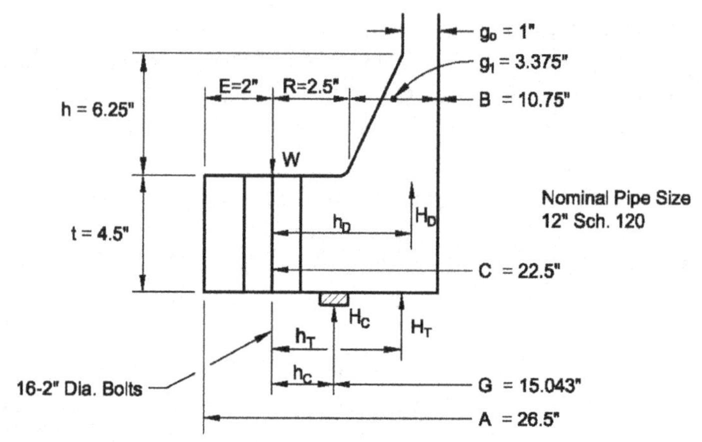

Figure 4.10 Sample calculation for a bolted flange (1 of 4).

Manufacturers like Taylor Forge provide manuals devoted to the design of flanges and include a one-page summary that includes step-by-step calculation sheets

An example calculation for a weld neck flange is shown in Figure 4.10

Sample Calculations for a Bolted Flange

Problem

What is the minimum required flange thickness of a weld neck flange with the following flange and gasket details and design conditions? Appendix 2 rules are to be followed to design the flange.

Design pressure, p = 2,500 psi

Design temperature = 250°F

Bolt-up and gasket seating temperature = 70°F

Flange material SA 105

Bolting material SA 325 Grade 1

No corrosion

Allowable bolt stress at design = S_b = 20,200 psi
Allowable flange stress at bolt-up and design = S_f = 17.500 psi
Gasket details: spiral-wound metal, fiber filled, stainless steel, inside diameter = 13.75 inches and width, N, is 1 inch
Gasket dimensions:

$b_0 = N/2 = 0.5$ inch and $b = 0.5\sqrt{b_0} = 0.3535$ inch

(Appendix 2, Table 2-5.2)

$G = 13.75 + (2 \times 1) - (2 \times 0.3535) = 15.043$ inches

Solution

1. Determine bolt loadings and size of bolts.

$N = 1;\ b = 0.3535;\ y = 10.000;\ m = 3.0$

(from Appendix 2. Table 2-5.1)

$H = \pi G^2 p/4 = \pi(15.043)^2(2500)/4 = 444,320$

(hydrostatic force)

$H_p = 2b\,\pi\,Gmp = 2(0.3535)\,\pi\,(15.043)(3.0)(2500) =$

250,590 (joint contact surface compression load)

$W_{m1} = H + Hp = (444,320) + (250.590) = 694,910$

(operating bolt load)

$W_{m2} = \pi bGy = \pi\,(0.3535)(15.043)(10,000)$

$= 167,060$ (gasket seating bolt load)

A_m = the greater of $W_{m1}/Sb = (694{,}910)/(20{,}200) =$ 34.4 inches^2 or $W_{m2}/Sa = (167{,}060)/(19{,}200) =$ 8.3 inches^2 (total required bolting cross-sectional area)

A_b = actual bolt area = 36.8 inch^2 for 16 bolts at 2-inch diameter (choose bolts in multiples of 4)

$W = 0.5(A_m + A_b)\,S_a = 0.5(34.4 + 36.8)\,(20{,}200) =$ 719,120 Ib (gasket seating)

$W = W_{M1} = 694{,}910$ Ib (operating condition)

2. Calculate the total flange moment for the design condition.

Flange Loads

$H_D = (\pi/4)\,B^2p = (\pi/4)(10.75)^2(2500) = 226{,}910$ Ib (hydrostatic end force based on I.D. of flange)

$H_G = H_p = 250{,}590$ Ib

$H_T = H - H_D = (444{,}320) - (226{,}910) = 217{,}410$ Ib

Lever Arms

$$h_D = R + 0.5_{g1} = \frac{C - B}{2} - g_1 + 0.5g_1$$

$$= \frac{22.5 - 10.75}{2} - 0.5 \times 3.375 = 4.1875\ \textit{inch}$$

$h_G = 0.5(C - G) = 0.5(22.5 - 15.043) = 3.7285$ inch

$h_T = 0.5(R + g_1 + h_G) = 0.5(2.5 + 3.375 + 3.7285)$ = 4.8018 inches (see Appendix 2, Table 2-6)

Flange Moments

$$M_D = H_D \times h_D = (226{,}910)(4.1875) = 950{,}190 \text{ in-lb}$$

$$M_G = H_G \times h_G = (250{,}590)(3.7285) = 934{,}320 \text{ in-lb}$$

$$M_T = H_T \times h_T = (217{,}410)(4.8018) = 1{,}043{,}960 \text{ in-lb}$$

$$M = M_D + M_G + M_T = 2{,}928{,}470 \text{ in-lb}$$

3. Calculate the total flange moment for bdt-up condition.

Flange Load at Sort Up

$$H_G = W = 719{,}120 \text{ Ib}$$

Lever Arm

$$h_G = 0.5(C - G) = 3.7285 \text{ inches}$$

Flange Moment

$$M_{\text{bolt up}} = H_G \times h_G = (719{,}120)(3.7285)$$
$$= 2,681,240 \text{ in-lb}$$

4. M_0 = greater of M_{design} or $M_{bolt\ up}$ (S_h/S_c) = 2,928,470 in-lb
5. Shape constants for flange:

$$K = A/B = (26.5)/(10.75) = 2.465$$

From Appendix 2, Figure 2-7.1: T = 1.35; Z = 1.39; Y = 2.29; U = 2.51:

$$g_1/g_0 = (3.375)/(1.0) = 3.375$$

$$h_0 = \sqrt{Bg_0} = \sqrt{(10.75)(1.0)} = 3.279$$

$$h/h_0 = (6.25)/(3.279) = 1.906$$

From App. 2, Figure 2-7.2: F = 0.57
From App. 2, Figure 2-7.3: V = 0.04
From App. 2, Figure 2-7.6: f = 1.0

$$e = F/h_0 = (0.57)/(3.279) = 0.1738$$

$$d = (U/V)h_0 g_0^2 = (2.51/0.04)(3.279)(1)^2 = 205.76$$

6. Calculation of stresses.
   Assume t = 4.5 inches:

$$L = (te + 1)/T + t^3/d = (1.3201) + (0.4429)$$
$$= 1.7630$$

Longitudinal hub stress:

$$S_H = (M_0/Lg_1^2 B = (1)(2{,}928{,}470)/(1.7630)(3.375)^2$$
$$\times (10.75)$$
$$S_H = 13.570 \text{ psi}$$

Radial flange stress:

$$S_R = (4/3te + 1)\,M_0/Lt^2 B = (2.0428)$$
$$\times (2{,}928{,}470)/(1.763)(4.5)^2(10.75)$$
$$S_R = 15{,}590 \text{ psi}$$

Tangential flange stress:

$$S_T = (YMo/t^2 B) - ZS_R = (2.29)(2.928.470)/(4.5)^2$$
$$\times (10.75) - 1.39(15{,}590)$$
$$S_T = 9{,}140 \text{ psi}$$

7. Allowable stresses:

$S_H \leq 1.5\ S_f:\ 13{,}570 < (1.5)(17.500) = 26{,}250$ psi; o.k.

$S_R \leq S_f:\ 15{,}590 < 17{,}500$ psi; o.k.

$S_T \leq S_f:\ 9{,}140 < 17{,}500$ psi; o.k.

$0.5\ (S_H + S_R) \leq S_f:\ 14{,}580 < 17{,}500$ psi; o.k.

$0.5\ (S_H + S_T) \leq S_f:\ 11{,}350 < 17{,}500$ psi; o.k.

Because all calculated stresses are below the allowable stresses, the selection of t = 4.5 inches is adequate. If an optimum minimum thickness of flange is desired, calculations must be repeated with a lesser value of t until one of the calculated stresses or stress combinations is approximately equal to the allowable stress even though other calculated stresses are less than their allowable stress.

Blind Flanges

Covered in Code Paragraph UG-34

Computations determine the thickness required for both operating conditions and gasket seating (Equation UW-34)

The more severe of the following are selected

Operating conditions

$$t = d\left(\frac{1.3P}{SE} + \frac{1.9Wh_G}{SEd^3}\right)^{0.5} \tag{18}$$

Gasket conditions

$$t = G\left(\frac{1.9W_a h_G}{S_a E G^3}\right) \tag{19}$$

MINIMUM THICKNESS OF PLATE		
Material	**Minimum Thickness**	**Code Reference**
Carbon and low-alloy steel	3/32 in. for shells and heads used for compressed air, steam, and water service	Par. UG-16 Par. UCS-25
	Minimum thickness of shells and heads after forming shall be 1/16 in. plus corrosion allowance	Par. UG-16
	1/4 in. for unfired steam boilers	
Heat-treated steel	1/4 in. for heat-treated steel	Par. UHT-16
Clad vessels	Same as for carbon and low-alloy steel based on total thickness for clad construction and the base plate thickness for applied-lining construction	Par. UCL-20
High-alloy steel	3/32 in. for corrosive service 1/16 in. for noncorrosive service	Par. UHA-20
Nonferrous materials	1/16 in. for the welded construction in noncorrosive service 3/32 in. for welded construction in corrosive service	Par. UNF-16
	Single-welded butt joint without use of backing strip may be used only for circumferential joints not over 24 in. outside diameter and material not over 5/8 in. thick	Table UW-12
Low-temperature vessels: 5, 8, 9 percent nickel steel	3/16 in.	Par. ULT-16
Cast-iron dual metal cylinders	5 in. (Max. diameter: 36 In.)	Par. UCL-29

Figure 4.11 Minimum thickness of plate required by the ASME code.

Minimum Wall Thickness and Nominal Plate Sizes

Minimum Thickness

Code permits various minimum thicknesses of plate (refer to Figure 4.11)

All vessels, regardless of alloy content, should have a minimum thickness of 3/16 inch, in addition to a 1/16-inch minimum corrosion allowance

In effect, this sets the minimum wall thickness required for pressure at 1/8 inch

Utilization of Available Metal and Commercial Thickness Plate

Once the minimum thickness has been determined, including corrosion, vessels are often brought to the next highest commercial plate thickness

Extra plate thickness is called corrosion allowances

Increasing the allowance of a vessel might necessitate redesign of several other pieces of equipment in its system in order to prevent the other pieces from limiting the system

The ability to prevent corrosion rate is not as precise as the ability to calculate pressure and structural requirements

Design of Welded Joints

Overall Considerations

The choice of weld design depends on a number of factors

The conditions of welding and accessibility have a definite influence on the type of weld joint; in small diameter vessels, the openings may have weld joints that are not accessible on both sides

The ASME Code itself limits the design and use of weld joints based on service, material, and location of weld

Company practices: certain types of welds have been found through experience to be more reliable and less prone to failure than others; for example, the use of backup rings in circumferential welds leads to many problems, yet the code allows this; many companies have found the following problems with backup rings:

1. They are difficult to check for weld quality by radiography
2. They invariably contain root weld defects

3. They greatly increase susceptibility to some form of fracture or cracking
4. They are not worth the potential savings

Terminology and Determination of Joint Efficiency

Code bases the joint efficiencies required in wall thickness calculations on the following factors:

Joint category based on the location of weld joint in the vessel, such as a girth weld

Joint type or how the weld was made, i.e. Double V, or Single V groove weld.

Location of the weld on the vessel

Figure 4.12 shows these four main categories

Joint type defines the configuration of a welded joint and not its location on the vessel

Figure 4.13 shows an example of a Type 2 weld joint

Figure 4.14 shows the various code weld types

Figure 4.15 shows the code limitations

Once the weld joint category and joint type are established, then the joint efficiency can be determined remembering that joint efficiency is also a function of the degree of radiography used to inspect the welds

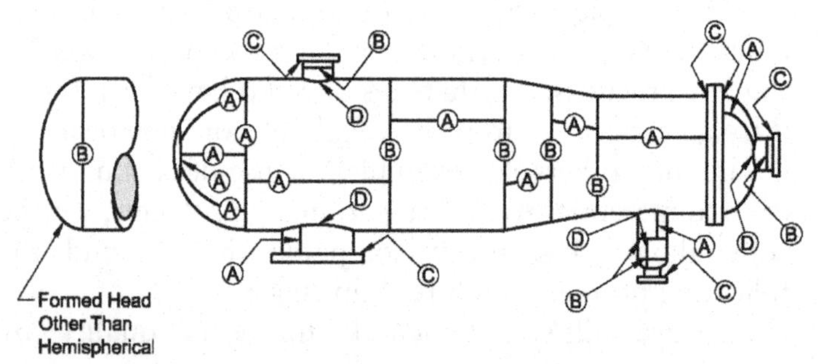

Figure 4.12 Categories of welded joints (see Figure 4.15 for special requirements based on service, material, thickness, and other design conditions).

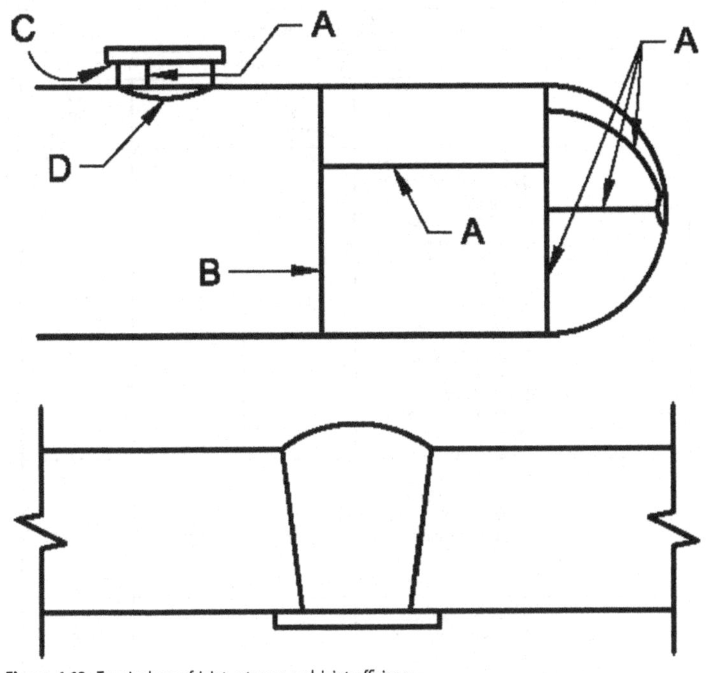

Figure 4.13 Terminology of joint category and joint efficiency.

Joint category defines the location of a joint in a vessel and does not define the type of joint.

Categories of Joints

A. Longitudinal welds in the main shell or attached nozzles; welds in a sphere, head, or flat-sided vessel; circumferential joints connecting hemi-heads to a vessel or part of a vessel
B. Circumferential welds in a shell or nozzle or connecting heads other than hemi-heads or part of a vessel

TYPES OF WELDED JOINTS			
	Joint Effeciency, E When The Joint Is:		
Types Code UW - 12	A. Fully Radiographed	B. Spot Examined	C. Not Examined
1 Butt Joints As Attained By Double-Welding Or By Other Means Which Will Obtain The Same Quality Of Deposited Weld Metal On The Inside And Outside Weld Surface. Backing Strip If Used Shall Be Removed After Completion Of Weld. Not Recommended Since They Are Crack Starters.	1.00	0.85	0.70
2 Single-Welded Butt Joint With Backing Strip Which Remains In Place After welding For Circumferential Joint Only	0.90	0.80	0.65
3 Single-Welded Butt Joint Without Use Of Backing Strip	-	-	0.60
4 Double-Full Fillet Lap Joint	-	-	0.55
5 Single-Full Fillet Lap Joint	-	-	0.50
6 Single-Full Fillet Lap Joint Without Plug Welds	-	-	0.45

Figure 4.14 Types of welded joints.

C. Welds connecting flanges, tube sheets, or flat heads to a vessel or part of a vessel and welds connecting flat-sided vessels
D. Welds connecting nozzles to a vessel or part of a vessel

	DESIGN CONDITION	WELD TYPE	RADIOGRAPHIC EXAM.	JOINT EFFICIENCY	POST-WELD HEAT TREATMENT
1	THE DESIGN OF PRESSURE VESSEL IS BASED ON JOINT EFFICIENCY 1.0 CODE UW-11(A)(5)	JOINTS A AND D. JOINTS A AND B (GIRTH SEAMS ONLY) SHALL BE TYPE NO. (1) AND NO. (2) JOINTS A AND C BUTT WELDS, UM-11(A)(5)(B) JOINTS A AND C BUTT WELDS IN NOZZELS AND COMMUNICATING CHAMBERS THAT WEITHER EXCEED 10 IN. NON PIPE SIZE NOR 1-1/8 IN. WALL THICKNESS DO NOT REQUIRE ANY RADIOGRAPHIC EXAMINATION EXCEPT AS REQUIRED FOR FERRITIC STEEL WITH TENSILE PROPERTIES ENHANCED BY HEAT TREATMENT UWT-57	FULL FULL, SPOT, NO PARTIAL THE ENTIRE WELD LENGTH REPRESENTED BY THIS PARTIAL RADIOGRAPH IS ACCEPTABLE UW-51 (C) (1)	1.0 TYPE (1) TYPE (2) 1.0 0.90 0.85 0.80 0.70 0.65 0.85 TYPE (1) 0.80 TYPE (2) EFFICIENCY OF GRITH SEAM MAY GOVERN ONLY WHEN SUPLEMENTARY LOADINGS AS MIND ETC. CAUSING LONGITUDINAL BENDING OR TENSION IN JUNCTIONS WITH INTERNAL PRESSURE	PER CODE UCS-56
2	PRESSURE VESSELS WHEN FULL RADIOGRAPHIC EXAMINATION IS NOT MANDATORY UW-11(B)	ALL BUTT WELDED JOINTS SHALL BE TYPE NO. (1) OR NO. (2) UM-11 (B)	BUILT WELDED JOINTS SPOT EXAMINED. UW-12 (B) SEAMLESS VESSEL SECTION AND HEADS WITH JOINTS B,C OR D SHALL BE DESIGNED FOR CIRCUMFERENCIAL STRESS USING A STRESS VALUE OF MATERIAL 85%	0.85 TYPE (1) 0.80 TYPE (2)	PER CODE UCS-56
3	THE VESSELS IS DESIGN FOR EXTERNAL PRESSURE ONLY, OR THE DESIGN OS BASED ON UW-12 (C) (SEE TABLE)	ANY WELDED JOINTS UW-11 (C)	NO RADIOGRAPHIC EXAMINATION IS REQUIRED	0.70 TYPE (1), 0.65 TYPE (2), 0.60 TYPE (3), 0.55 TYPE (4), 0.50 TYPE (5), 0.45 TYPE (6) IN ALL OTHER DESIGN CALCULATION 80% OF STRESS VALUE OF MATERIAL SHALL BE USED	PER CODE UCS-56
4	VESSELS CONTAINING LETHAL SUBSTANCES UW-2(A)	JOINTS A SHALL BE TYPE NO. (1), UM-2(A) (1) (A). JOINTS B AND C SHALL BE TYPE NO. (1) OR TYPE NO. (2) UW-2 (A) (1) (B) JOINTS D SHALL BE FULL PENETRATION WELDS EXTENDING THROUGH THE ENTIRE THICKNESS OF THE VESSELS OR NOZZEL WALL UW-2 (A) (1) (C)	FULL ALL BUTT WELDED JOINTS IN SHELL AND HEADS SHALL BE FULLY RADIOGRAPHED EXCEPT EXCHANGER TUBES AND EXCHANGER. UW-2 (A) AND (3) AND PER UW-11 (A) (4)	1.0 1.0 0.9* * TO BE USED FOR LONGITUDINAL STRESS CALCULATION (GIRTH SEAMS)	VESSELS FABRICATED OF CARBON OR LOW ALLOY STEEL SHALL BE POST-WELD HEAT TREATED UW-2C
5	VESSELS OPERATED BELOW -20°F AND IMPACT TEST IS REQUIRED FOR THE MATERIAL OR WELD METAL. UW-2 (B)	JOINTS A SHALL BE TYPE NO. (1) (EXCEPT FOR TYPE 304 STAINLESS STEEL) JOINTS B SHALL BE TYPE NO. (1) OR NO. (2), UW-2 (B) (1) AND (2). JOINTS C FULL PENETRATION WELDS EXTENDING THROUGH THE ENTIRE SECTION OF THE JOINT. JOINTS D FULL PENETRATION WELDS EXTENDING THROUGH THE ENTIRE SECTION OF THE JOINT. UW-2(B)(2) AND (3)	FULL SPOT NO	TYPE (1) TYPE (2) 1.0 0.90 0.85 0.80 0.70 0.65	PER CODE UCS-56
6	UNFIRED STEAM BOILERS WITH DESIGN PRESSURE EXCEEDING 50 PSI	JOINTS A SHALL BE TYPE NO. (1) JOINTS B SHALL BE TYPE NO. (1) OR (2) WHEN THE UW-2 (C)	ALL BUTT WELDED JOINTS IN SHELL AND HEADS SHALL BE FULLY RADIOGRAPHED EXCEPT UNDER THE PROVISIONS OF UW-11 (A)(4) UW-2 (C)	1.0 1.0 TYPE (1) 0.9 TYPE (2)	VESSELS FABRICATED OF CARBON OR LOW ALOY STEEL SHALL BE POST-WELD HEAT TREATED UW-2 (C)
7	PRESSURE VESSELS SUBJECT TO DIRECT FIRING	JOINTS A SHALL BE TYPE NO. (1) JOINTS B SHALL BE TYPE NO. (1) OR NO. (2) WHEN THE THICKNESS EXCEEDS 5/8 IN. UW-2 (D)	FULL SPOT NO	TYPE (1) TYPE (2) 1.0 0.90 0.85 0.80 0.70 0.65	WHEN THE THICKNESS AT HELDED JOINTS OF CARBON STEELS (P-NO. 1) EXCEED 5/8 IN. AND ALL THICKNESS FOR LOW ALLOY STEELS (OTHER THAN P-NO. 11 POST-WELD HEAT TREATMENT IS MANDATORY
8	ELECTROSLAG WELDING	ALL BUTT WELDS UW-11 (A) (6) ANY WELDS	FULL IN LIEU OF RADIOGRAPHING ULTRASONIC EXAMINATION	1.0 TYPE (1) 0.9 TYPE (2)	PEN CODE UCS-56
9	FINAL CLOSURE OF VESSELS	ANY WELDS UW-11 (A)(7)	FULL ULTRASONIC EXAMINATION WHEN THE CONSTRUCTION DOES NOT PERMIT RADIOGRAPHS	1.0 TYPE (1) 0.9 TYPE (2)	PEN CODE UCS-56

Figure 4.15 Design of welded joints.

Note: A butt weld is defined as a weld connecting two pieces of material with surfaces that are 30 degrees or less from being in the same plane.

Types of Joints

Type defines the configurations of a welded joint; it does not define the location of the joint in the vessel—for example, a Type 2 joint is a single-sided butt weld with a backing strip left in place

Joint Efficiency

The joint efficiency (E) or percentage of stress value (S) to be used in calculating required thickness or maximum allowable working pressure is a function of the following:

Type of welded joint

Degree of radiography performed

5 Fabrication, Welding, and In-Shop Inspection

▶ OVERVIEW

This chapter discusses the fabrication of pressure vessels:

Plate materials

Methods of forming the shell and head components and nozzles

Fabrication welds

Welding processes and procedures

In-shop inspection

The fabrication of a pressure vessel requires the following:

A thorough knowledge of materials, welding, and NDE technologies

Materials

Welding

Non Destructive Examination technologies

A high level of skilled workmanship

Figure 5.1 schematically illustrates the typical sequence of operations that must be performed to fabricate a pressure vessel

This section covers the following points:

Describes the various operations involved in vessel fabrication

Provides the designer/purchaser with a general understanding of how vessels are fabricated

Does not provide instructions for the fabrication of a pressure vessel

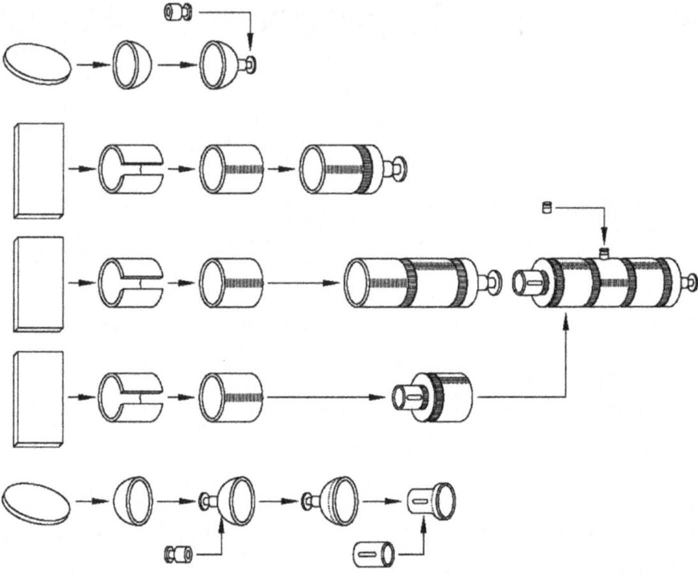

Figure 5.1 Typical sequence of operations for fabrication of a pressure vessel.

▸ PLATE MATERIALS

Overview

Plate materials are procured according to the materials specification and grade shown on the vessel drawing or proposed by the fabricator

Materials are selected to obtain the properties required for the service conditions and process environment:

- Strength
- Hardness
- Toughness
- Corrosion resistance
- Designated by reference to a "SA Specification" in the ASME Code

ASME Code Requirements

Fabricator orders the plate material using a *purchase specification*

Part of the quality control system

Required to obtain ASME Code certification of authorization for "U" and "UM" symbols

Materials must comply with SA specification in ASME Code, Section II

Supplier must provide the fabricator with a "materials test report" certifying that all the requirements of the specification have been met

Authorized inspector examines the test reports:

- Verifies that the materials test reports represent the lots of material delivered to the fabricator
- Verifies that the requirements of ASME Code, Section II, have been met

Fabricator Considerations

Fabricator must recognize that forming, welding, and the post-weld heat treating (PWHT) can significantly alter the properties of the material after it has been received from the supplier

Fabricator must consider the changes in materials properties that will result from the fabrication procedure that will be used and must prepare the materials purchase specification to allow for these changes

Fabricator may perform supplemental tests required to confirm that the materials properties specified in the pressure vessel specifications will be obtained in the completed vessel; this can be done in two ways:

- By simulating the anticipated thermal history with test coupons

By having the materials supplier perform the tests
Code does not require supplemental testing in all cases, thus the company inspector should review the test data
Deformation of plate material associated with the forming of the shell and head components reduce the thickness of the plate and thus require plate material to be thicker than the minimum required thickness (plus corrosion allowance); the fabricator should account for this

FORMING OF SHELL AND HEAD COMPONENTS

General Considerations

Shell and head components are usually formed by the fabrication from plate materials that are procured from materials suppliers
Forming causes deformation that can reduce the thickness of the plate
This reduction in thickness must be considered by the fabricator when the plate material is procured, to make certain that the minimum required thickness plus corrosion allowance is obtained

Cylindrical Shell Ring Courses

Fabricated by "rolling and welding" process
Plate with the required thickness is cut to the size required for the diameter of the vessel and the length of the ring course
Large-diameter or thick-walled vessels may require two or more plates to obtain the circumference, and consequently they may require two or more longitudinal welds

Weld Bevel Preparation

Machines on the plate edges after the plate is cut to size must remove all shear-damaged material

Oxy-fuel gas or plasma-arc cutting used in conjunction with cutting the plate to size

Reduces time and cost of machining

May require finish machining or grinding to make surface acceptable for welding and remove thermally damaged material

Rolling of Plate into Shell Ring Course

Refer to Figure 5.2

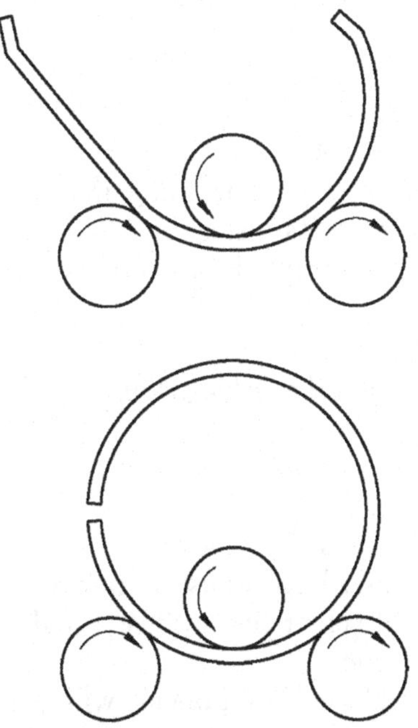

Figure 5.2 Rolling of plate into a cylindrical shell ring course.

Correct diameter is obtained when the edges of the rolled plate come together to form the joint preparation for the longitudinal weld

"Peaking" at the longitudinal weld is minimized by "crimping" of the plate edges prior to rolling

Code Paragraphs UG-80 and AF 130 require the roundness of a rolled cylinder be 1% of the nominal diameter for which the cylindrical component was designed

Effect of Plate Thickness

Plates less than 3 inches thick are generally cold rolled

Plates greater than 3 inches thick may have to be hot rolled to reduce the force required to form the plate into a cylinder

Stress-Relief Heat Treatment

Recommended after cold forming if the fiber elongation exceeds 5% for carbon steel or 3% for low-alloy Cr-Mo steel; Code Paragraph UCS-79 provides the following equation:

$$\%\ Extreme\ Fiber\ Elongation = \frac{50t}{Rt}\left(1 - \frac{R_f}{R_0}\right) \quad (1)$$

Where

t = Plate thickness, in inches

R_f = Final center line radius, in inches

R_0 = Original center line radius (equals infinity for flat plate), in inches

Rolled cylindrical shell components will rarely exceed these limits

Effects of Hot Rolling at High Temperatures

Can cause grain growth in plate materials that were given a normalizing heat treatment by the supplier

Can reduce the C_V impact toughness at temperatures above 1750°F and may have to be heat normalized after rolling to restore the minimum C_V-impact toughness

Fabrication by the "ring forging" process

Produces a ring course of the required diameter and thickness without the need for longitudinal welds

Only ring courses with thickness of 4 inches and larger are manufactured in this manner

Heads

"Pressing" of Formed Heads

Elliptical and spherical heads formed from one piece of plate

Plate of the required thickness is cut out to a circular shape with a diameter large enough to provide all the material required for the formed head

Circular plate is then pressed between mating dyes that have been manufactured with the inside and outside dimensions and contour of the head (refer to Figure 5.3)

Forming of a head reduces the thickness of the plate; the fabricator must account for this when procuring the materials

It is acceptable to weld two or more pieces of plate together when one piece is not large enough for the required diameter

Weld should receive full-coverage Radiograph Inspection (RT)

Knuckle of elliptical and for spherical heads should be Visual Inspection (VT) wherever the deformation of the plate associated with forming is not severe

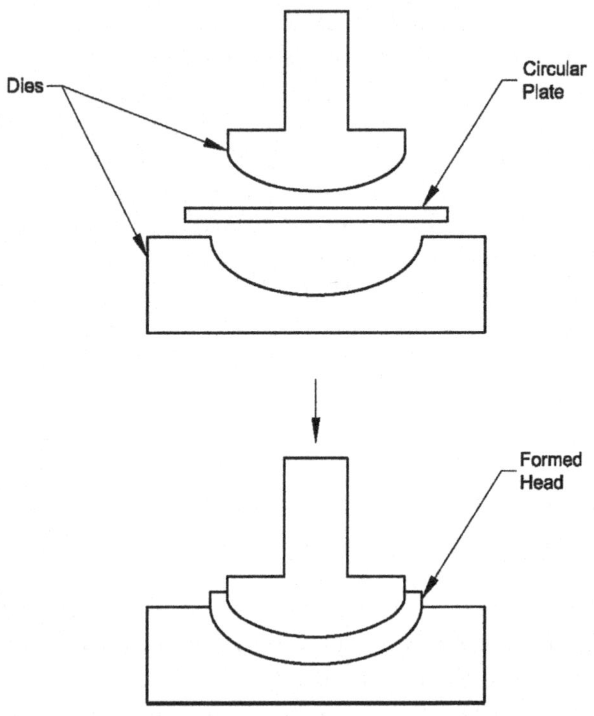

Figure 5.3 Pressing of formed heads.

Hot and Cold Rolled

Similar considerations as shell ring courses

Code Paragraph UCS-79 provides the following more restrictive formula for calculating the fiber elongation in formed heads (for double curvature):

$$\% \, Extreme \, Fiber \, Elongation = \frac{75t}{R_f}\left(1 - \frac{R_f}{R_0}\right) \quad (2)$$

Formed heads may require a stress-relief heat treatment when cylindrical components in the same vessel do not

PWHT requirements of the code can also serve the function of the stress relief required by the fiber elongation

The fabricator can purchase formed heads from a vendor, but the fabricator can also form them from plate material

Heads can be fabricated from several preshaped pieces of plate, called *core sections*; each piece is formed to the required contour of the head, and they are then welded together

▶ NOZZLES

Integrally Reinforced Nozzles

Recommendations for Shell Components

2 inches thick and larger

Maximum design temperature ≥650°F

Nozzles May Be Subjected to NDE

To meet the higher-quality standards desired for service at higher pressures and temperatures

To detect flaws that might develop during service

Nozzle Forgings

All forgings should be forged as close as practicable to finished size and shape to develop metal flow in the direction most favorably oriented for resisting the stresses encountered in service

The metal flow in forgings that are forged close to the finished shape follows the surface contour, as shown in Figure 5.4A

Some forging vendors produce a forged billet and then machine the billet to the finished shape; the metal flow does not follow the surface contours (Figure 4B) and thus will not be favorably oriented for the bending stresses that can develop

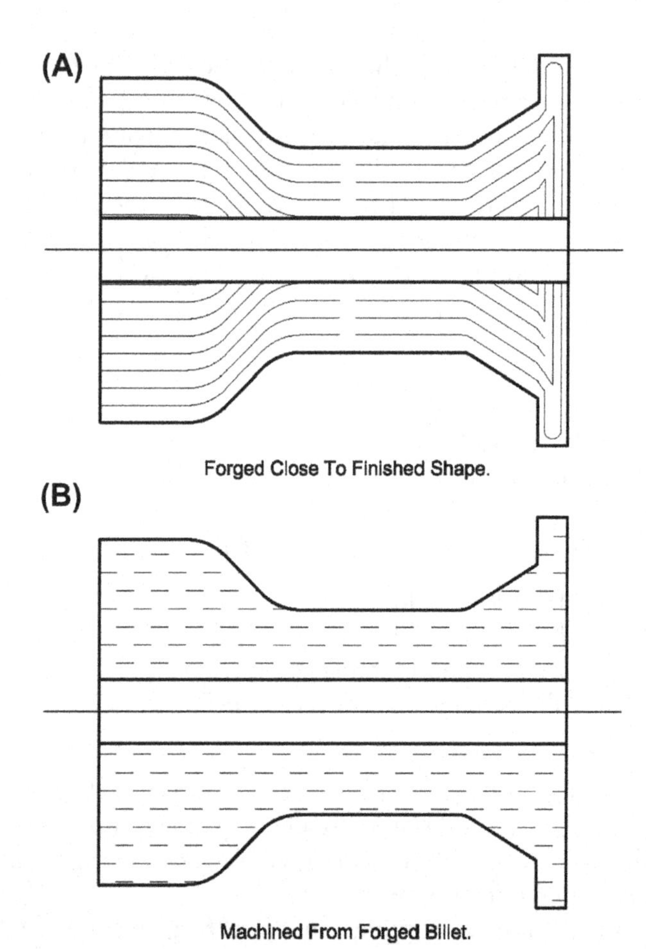

Figure 5.4 Integrally reinforced nozzle forgings.

Built-Up Nozzles

General Considerations

Nozzles can be fabricated from

Standard size pipe and ANSI piping flanges

Rolled-and-welded plate with plate flanges

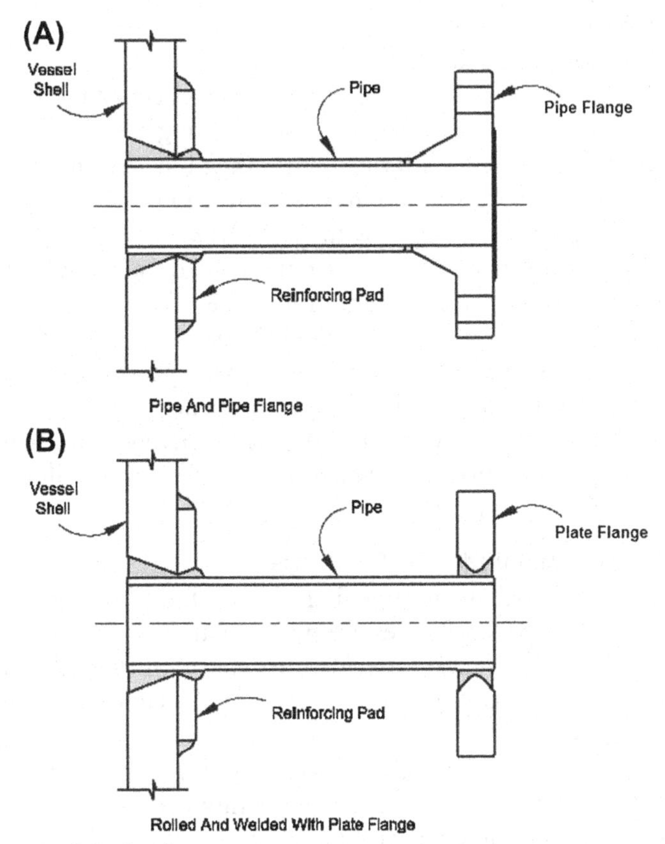

Figure 5.5 "Built-up" nozzles.

Refer to Figure 5.5
These types of nozzles are satisfactory:
Shell components less than 20 inches thick and
When maximum design temperatures are less than 650°F
Usually less expensive to fabricate than integrally forged nozzles

Pipe and Pipe Flange Nozzles

Fabrication must do the following:
- Procure the pipe and forged pipe flanges according to the materials specified in the vessel drawing
- Incorporate any additional specification requirements into the materials purchase specification

Pipe flanges must comply with ANSI B16.5

Full-penetration welds are recommended, although the code permits partial penetration welds

Partial penetration welds are undesirable

Act as crack initiators as locations with potentially high complex stresses and bending movements

More difficult to examine nondestructively for flaws during fabrication or for flaws that might develop during service

Rolled and Welded Nozzles with Plate Flanges

Used in larger-diameter openings where standard pipe sizes and ANSI B16.5 flanges are not available

The plate for rolled and welded nozzles with plate flanges must be procured like the plate for cylindrical shell components

Fabrication of the cylindrical nozzle barrel is similar to the fabrication of a cylindrical shell component

Only full-penetration welds should be used for this type of nozzle, as for nozzles built up from pipe and pipe flanges

► FABRICATION WELDS

Longitudinal Welds

Referred to as Category A by the ASME Code

Weld joints with the highest design stresses in the entire vessel

Full-penetration welds that are welded from both sides of the shell (referred to as Type 1 in the ASME Code) provide the highest quality weld joint

Welding

The plate materials for cylindrical shell components are rolled (formed) into cylindrical shapes with the required diameters, and the bevels (joint preparations) for the longitudinal welds are made

Code Paragraphs UW-33 and AF-142 require misalignment of the mating joint preparations for longitudinal welds to be within the following limits:

SECTION THICKNESS (T). IN.	MISALIGNMENT TOLERANCES
Up to ½, incl.	¼t
Over ½ to ¾, incl.	⅛t
Over ¾ to 1½, incl.	⅛t
Over 1½ to 2, incl.	1/16t (3/8 in. max)

Longitudinal welds in cylindrical shell components are commonly made by automatic submerged arc welding (SAW) from either the outside or the inside of the shell, followed by back gouging and manual shielded arc welding (SMAW) from the opposite side to complete the weld (Figure 5.6)

The joint preparation is configured to perform most of the welding with the more efficient automatic SAW process, and SMAW from the opposite side is only employed to fill and cover the back gouger groove

The groove for the longitudinal weld is generally completely filled with SAW weld material before work is initiated on the inside surface; the first layer of SAW weld metal

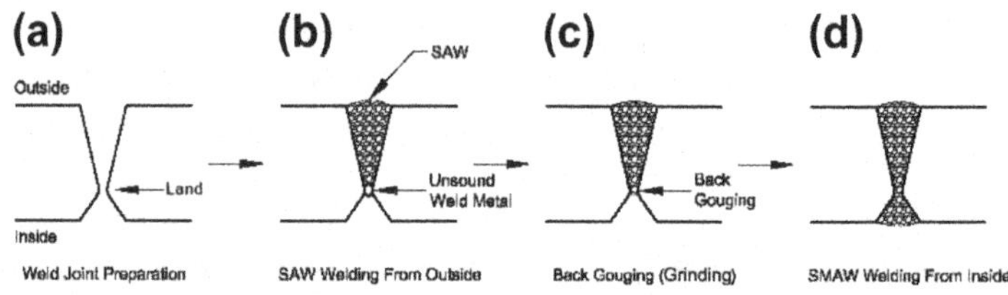

Figure 5.6 Typical longitudinal (Category A) weld in a cylindrical pressure vessel shell component.

deposited (root pass) generally contains considerable unsound materials; it must be removed by gouging or grinding prior to beginning manual welding (SMAW) from the inside

It is essential that all of the welding procedures used by the fabricator are properly qualified according to Section XI and that these procedures are strictly adhered to during production welding

Some distortion (or "peaking") can occur during welding of the longitudinal joint, and thus the shell may have to be re-rolled after welding to be within the code tolerances for out of roundness

NDE

Code Paragraphs UW-11 and UCS-57 require "RT" examination of longitudinal welds for shell components of types:

Carbon steel vessels exceeding 11/4-inch thick

Low-alloy 11/4 Cr-1 Mo vessels exceeding 5/8 inches

All low-alloy 21/4 Cr-1 Mo vessels regardless of thickness

Spot RT or VT (i.e., no RT) is permitted for components of vessels having thicknesses that are thinner than these limits, provided reduced weld joint efficiencies are used for the design (Code Paragraph UW-12)

As a minimum, spot RT is recommended

Full RT examination is recommended for the following conditions:

1. For services at low temperatures when undetected flaws could cause brittle fracture
2. For services when cyclic pressures or temperatures are encountered and undetected flaws could initiate fatigue cracks

ASME Code, Section VIII, Division 2

Requires full RT of all longitudinal welds regardless of thickness

Authorized inspector is obligated to review all radiographs that are required by the code

Ultrasonic Testing examination is recommended for all longitudinal welds in shell components 3 inches thick or greater or when gas metal arc welding (GMAW) is employed, regardless of thickness

Nozzle Welds

General Considerations

Known as Category D welds

Can be the most difficult to make

More flaws are encountered than in longitudinal welds (Category A welds) and girth welds (Category B welds)

Subjected to relatively high and complex stresses, especially if piping connections apply external loads and moments

Welding

Dropouts (i.e., holes) are cut into the shell or head components of a vessel, usually using oxy-fuel gas cutting, and the weld bevels (joint preparations) are also oxy-fuel gas cut or machined

The nozzle barrel is then inserted into the opening, and the proper fit-up for welding is maintained by the use of clips welded to the shell component and nozzle

Nozzles can be welded into shell and head components either before or after the various components are joined together depending on the fabricator's shop facilities and most efficient plan for final assembly

Welding the nozzles into the shell or head component is generally accomplished manually (SMAW)

Automatic welding (SAW) is sometimes used for large-diameter nozzles in relatively thick shells

Some fabricators like to use semiautomatic welding (flux-cored arc welding [FCAW]), which can be more efficient than manual welding because of the continuous wire feed; however, a high incidence of underbead cracking can occur with FCAW unless low-hydrogen electrodes are used

When integrally reinforced nozzle forgings are used (see Figure 5.4), double-V joint preparations are generally employed to permit welding from both the inside and outside surfaces, similar to that described for the welding of longitudinal joints

When built-up nozzles with a reinforcing pad are used (see Figure 5.5), a single-V joint preparation is more commonly used for installation of the nozzle barrel, with all of the welding performed from the inside

The outside surface of the weld is subsequently gouged or ground to remove unsound weld metal, and the reinforcing pad is then welded to the nozzle barrel and shell component from the outside

The company inspector should determine that all of the specified tolerances for orientation and elevation have been obtained after the welding is completed, as some distortion can occur during welding

NDE

Division 1 does not require any NDE of nozzle welds (Category D welds), regardless of the extent of RT and the joint efficiency; the configuration of most nozzle welds (see Figure 5.5) usually does not permit satisfactory RT

Division 2 (paragraph AF-240) requires UT examination of these welds if RT cannot be satisfactorily performed

Girth Welds

General Considerations

The cylindrical shell and head components are joined together with girth welds (Category B welds)

These welds are usually made using a welding procedure employing a combination of automatic welding (SAW) and manual welding (SMAW) that is very similar to these described for longitudinal welds

Welding

The cylindrical shell (or head) components to be joined together are positioned horizontally on rolls that can revolve the mating components in unison for girth welding

Code Paragraphs UG-80 and AF-142 require obtaining the following tolerances for misalignment of girth welds:

SECTION THICKNESS (T). IN.	MISALIGNMENT TOLERANCES
Up to ½, incl.	¼ t
Over ½ to ¾, incl.	¼ t
Over ¾ to 1½, incl.	3/16 inch
Over ¾ to 1½, incl.	⅛ t
Over 1½ to 2, incl.	⅛ t
Over 2	⅛ t (2/3 max.)

Aforementioned tolerances are somewhat larger than required for longitudinal welds

Almost always necessary to use clips or strong backs (temporary attachments) welded to the mating components to obtain fit-up within these limits and to assure that they are maintained during welding

Most of the welding is performed by SAW from the outside or inside of the shell

The mating shell (or head) components being welded are tacked together and are then revolved in unison under the welding head, so the weld metal is deposited in the groove between the two components

The unsound (SAW) weld metal in the root pass is removed by back gouging or grinding, and the weld is completed by manual welding (SMAW) from the opposite side of the shell

The discussion for welding of longitudinal joints also applies to girth welding

NDE

Division 1

- Permits spot RT of girth welds (Category B welds) when full RT is required for longitudinal welds (Category A welds) for design with a joint efficiency of 1.0
- The justification is that the maximum stress transverse to a girth weld is normally half that in a longitudinal weld

Division 2

- Does not allow relaxation of the RT coverage of girth welds
- All other NDE requirements for girth welds are the same as those discussed for longitudinal welds

Skirt Attachment Weld

General Considerations

Attaching a support skirt to a pressure vessel does not involve a pressure-containing weld

Must provide the integrity required

Support the vessel, including wind and earthquake loads in a manner that does not jeopardize the pressure-containing integrity of the vessel

Usually fabricated by rolling and welding, similar to the formation of the cylindrical shell components of the vessel

Welding

Code Paragraphs UW-28(c) and AF-210.4 require that the welding procedure be qualified according to Section IV

Procedure should be similar to those used for pressure-containing welds in the vessel, including requirements for preheat and post-weld heat treatment

A support skirt can be directly welded to the bottom head of a pressure vessel, as shown in Figure 5.7A

Automatic welding (SAW) or manual welding (SMAW) is normally used to make this type of skirt attachment weld

Cannot be readily NDE, and flaws almost always exist in the "crotch," which could propagate by fatigue if they are subjected to thermal cycling

Attachment of the support skirt to the bottom head employing a weld pad as shown in Figure 5.7B provides a higher integrity attachment that is preferred for heavy and thick wall vessels or those that will be subjected to thermal cycling, which could initiate fatigue cracks

Weld pad can be deposited on the head by either manual welding (SMAW) or automatic welding (SAW)

Weld pad is subsequently machined to a smooth contour that incorporates a weld bevel for attachment of the skirt;

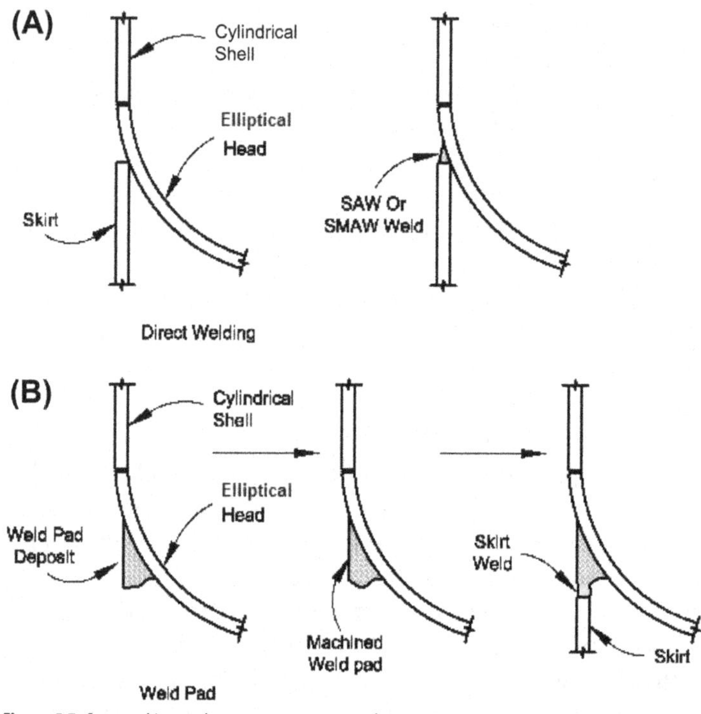

Figure 5.7 Support shirt attachment to a pressure vessel.

the skirt is then welded to the machined weld pad, and this joint preparation permits either RT or UT examination

Post-Weld Heat Treatment (PWHT)

General Considerations

PWHT is a very important factor that can have a significant effect on the integrity and reliability of the vessel

Omitting or improperly performing this function can lead to failure of the vessel

Attaching sufficient thermocouples to the vessel during PWHT is recommended to assure that the minimum temperature for PWHT required by the code has been obtained and the temperature should not vary more than 50°F over the entire vessel

Temperature variations during heating and cooling should not vary more than 150°F over the entire vessel, to prevent damage by the development of thermal stresses

PWHT can be performed locally for individual welds when a completed vessel is too large to fit inside of the heat-treating furnace

It is important to make certain that the temperature gradients resulting from the local heat treatment do not develop thermal stresses that damage the vessel

It is important that no welding is performed on pressure-containing components of a vessel after it is PWHT

Occasionally, by oversight, some external attachments to the vessel shell must be welded to the shell after the PWHT has been performed; if this occurs, a second PWHT is required

A company inspector should maintain sufficient surveillance to do the following:

- Assure that the PWHT is properly performed in accordance with the code
- Assure that no welding is performed after the vessel is heat treated

▸ WELDING PROCESSES AND PROCEDURES

Welding Processes

Shielded Metal Arc Welding (SMAW)

Referred to as "stick" welding

Strictly a manual process that uses a consumable flux-coated metal electrode to provide filler metal, flux, and slag (Figures 5.8 and 5.9)

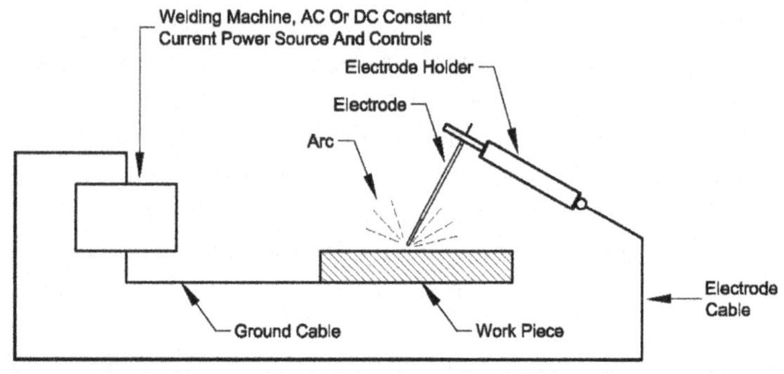

Figure 5.8 Typical welding circuit for a shielded metal arc welding (SMAW).

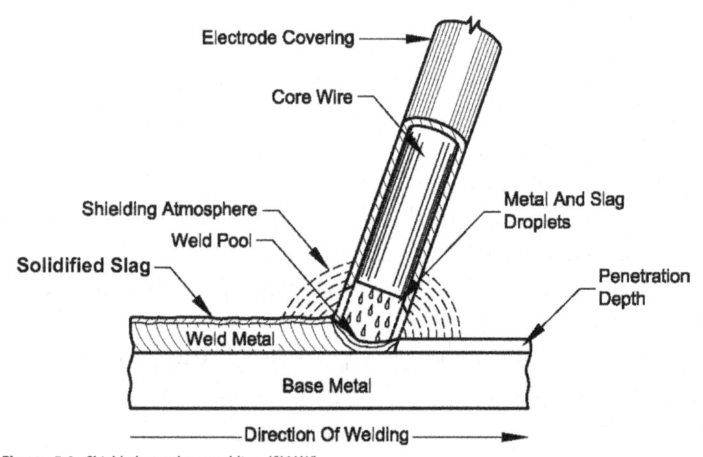

Figure 5.9 Shielded metal arc welding (SMAW).

Most versatile welding process used for pressure vessel fabrication

Can be used for all positions, thicknesses, and joint designs

Equally suited for shop or field fabrication

Process is preferred for pressure vessel nozzle and attachment welds
Acceptable for circumferential and longitudinal welds in lieu of the typically used SAW process

Gas Tungsten Arc Welding (GTAW)

Referred to as "TIG" welding
Utilizes a nonconsumable tungsten electrode and separate filler metal in the form of a wire (not used for thin sections)
Inert shielding gas is supplied through an annular nozzle around the tungsten electrode
Better suited for shop fabrication than field fabrication because air movement must be less than 5 mph to maintain the inert gas blanket
GTAW is predominantly a manual process, but automatic processes have been developed for high-quality pipe and tube welding (Figure 5.10)
Preferred for root passes of welds that cannot be back-welded, such as vessel-closure welds or welds on small-diameter sections
Used to weld very thin stainless-steel parts like vessel internals
Can be used for all positions, but welding confined joints may not be feasible because it requires both of the welder's hands
Manual process is slow and not economically attractive for thick welds or filler passes because of low deposition rates

Gas Metal Arc Welding (GMAW)

Referred to as "MIG" welding
Utilizes an automatically fed consumable electrode in the form of a wire from a spool for the filler metal

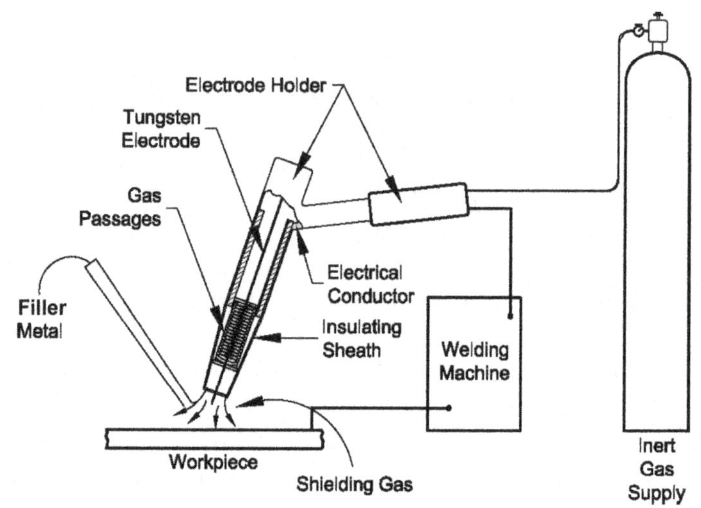

Figure 5.10 Equipment for a gas tungsten arc welding (GTAW).

Inert shielding gas is supplied through an annular nozzle at the contact tip of the gun

Process is better suited for shop fabrication because air movement must be less than 5 mph to maintain the inert gas blanket (Figures 5.11 and 5.12)

Predominantly a semiautomatic process, but automatic processes are sometimes used for weld overlays

Depending on the current, voltage (arc length) and shielding gas composition, three modes of metal transfer are commonly used for GMAW:

1. Short-circuiting transfer (also called *short arc* or *interrupted arc transfer*) (Figure 5.13)
2. Globular transfer (Figure 5.14)
3. Spray transfer (Figure 5.15)

Short-circuiting transfer GMAW is a low-heat input form of welding that can be used for all positions

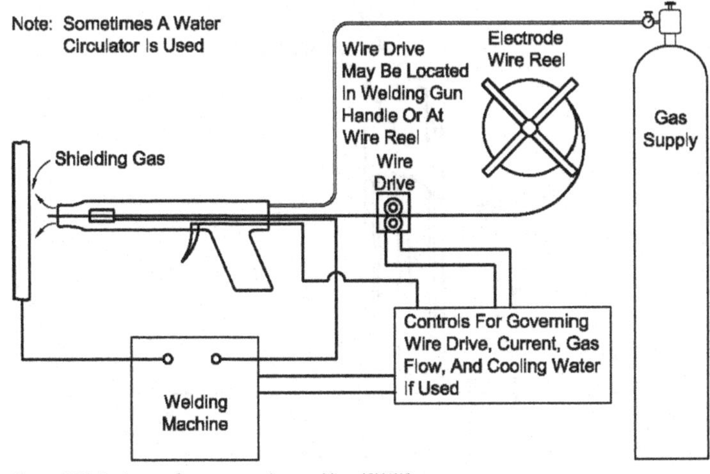

Figure 5.11 Equipment for a gas metal arc welding (GMAW).

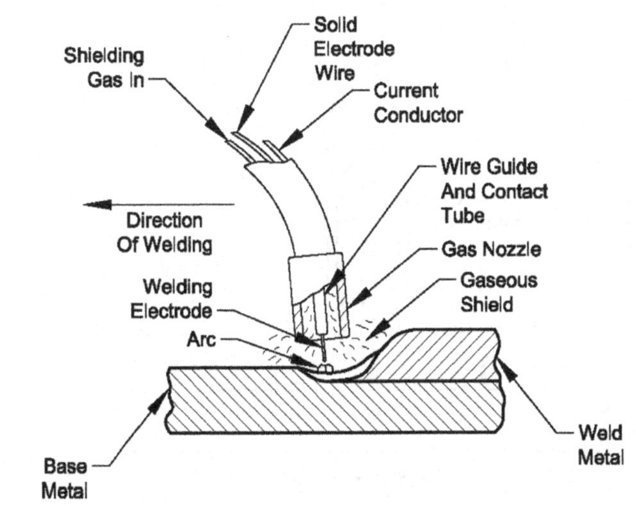

Figure 5.12 Gas metal arc welding process (GMAW).

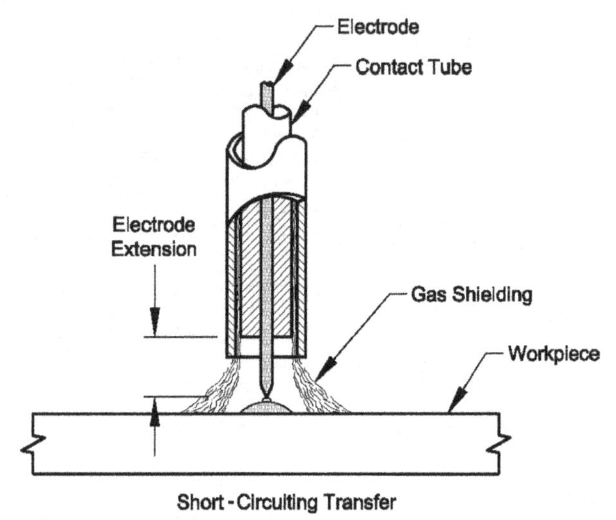

Figure 5.13 GMAW—short-circuiting transfer.

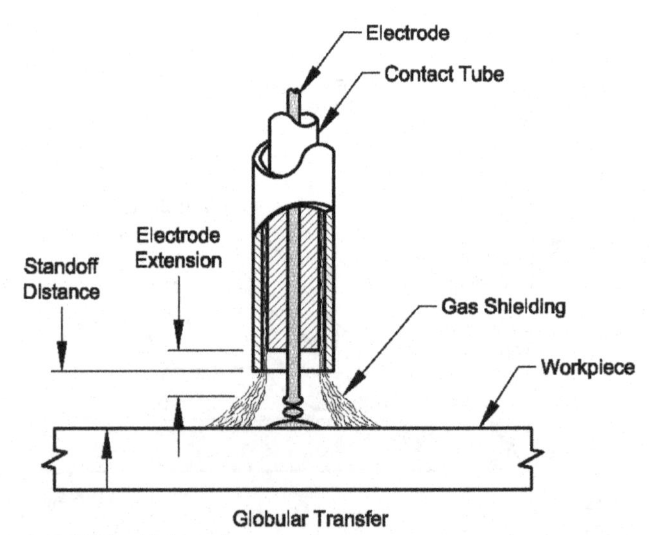

Figure 5.14 GMAW—globular transfer.

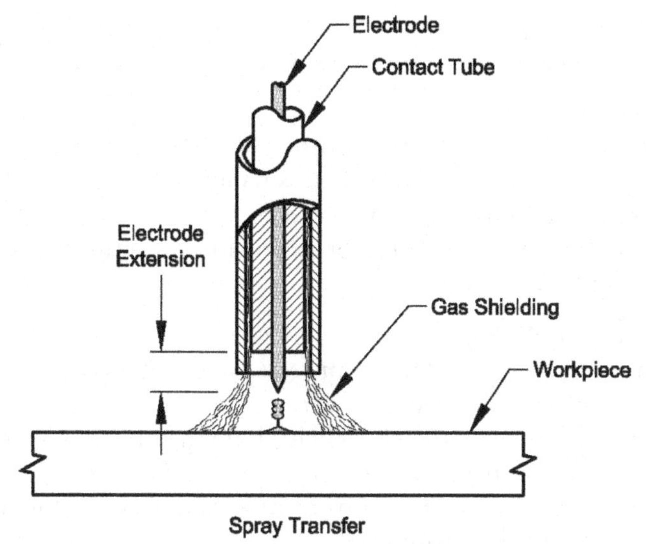

Figure 5.15 GMAW—spray transfer.

Process is notorious for producing lack-of-fusion defects

Normally not recommended for pressure vessel fabrication, except for the following applications:

- Root passes on circumferential, longitudinal, or nozzle-to-shell weld only if back gouged and back welded
- Root passes on circumferential piping welds for fabricated nozzles or internal piping
- Root passes on non-pressure-containing vessel internals

Globular and spray transfer GMAW processes are high-heat input processes, which are acceptable for pressure vessel fabrication

Drawbacks of the globular and spray transfer (GMAW) process:

- Generally limited to the flat position
- Typically used only for circumferential and longitudinal welds, rather than for nozzle and attachment welds

Must compete with the faster (SAW) process for welding applications and hence are not economical

Pulsed-Current GMAW

Modification of the spray transfer GMAW process

Utilizes a pulsed current and smaller diameter wire to achieve all-position capabilities

Typical applications are for corrosion-resistant overlays and nozzle welds

Flux-Cored Arc Welding (FCAW)

Variation of the (GMAW) process, which uses flux-cored wire instead of solid bare wire

The flux forms a slag, which helps hold the molten metal in place so the process can be used for all positions

Used primarily in the spray transfer mode, but globular transfer is not uncommon

Noted for high deposition rates

Applications include nozzle welds and welds to vessel attachments such as stiffening rings, insulation support rings, internals, and supports

Process can be either gas-shielded or self-shielded (with or without inert shielding gas) (Figures 5.16 and 5.17)

The gas-shielded FCAW process (FCAW-G), like the GMAW process, is much better suited for shop fabrication because air movement must be less than 5 mph to maintain the inert gas blanket

The self-shielded FCAW process (FCAW-SS) is equally suited for either shed or field fabrication

The FCAW-SS process should not be accepted for pressure vessel fabrication without special review by a materials or welding specialist on a case-by-case basis

This precaution should be taken because of the need for careful consumable selection and the significant training

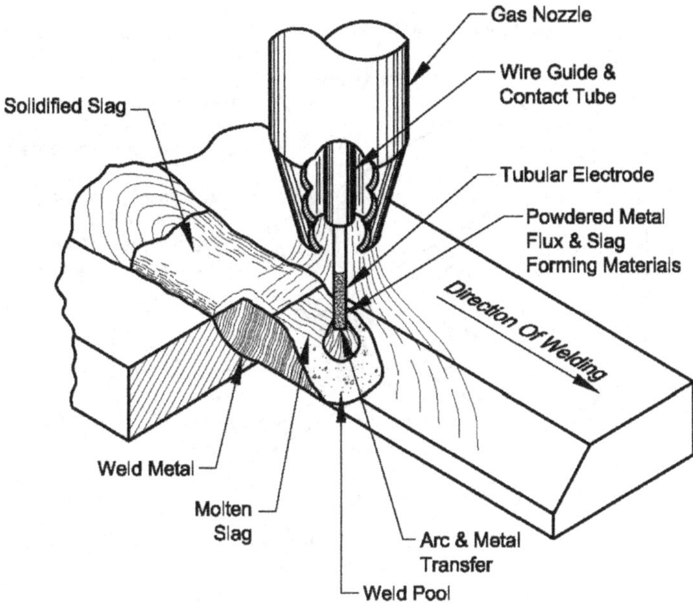

Figure 5.16 Gas-shielded, flux-cored arc welding (FCAW-G).

and supervision required for obtaining reliable welds with this process

The FCAW-G process is acceptable for pressure vessel fabrication only if special low-hydrogen, flux-cored wire is used

Submerged Arc Welding (SAW)

Uses a continuously fed consumable electrode (or electrons) in the form of a wire or strip (for weld overlays) from a coil and a granular flux

Process similar to GMAW except that the arc is submerged in a granular flux, which melts and provides shielding from the atmosphere

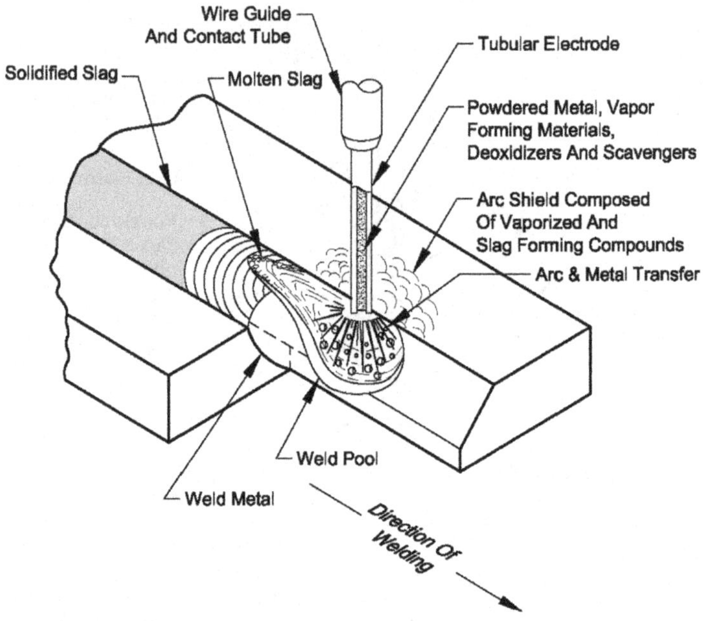

Figure 5.17 Self-shielded, flux-cored welding (FCAW-SS).

Because of the need to hold the flux in place until the molten metal cools, the process is almost always used in the flat position (Figure 5.18)

However, some specialized equipment has been developed for supporting the flux in the horizontal position to make girth seams on large field-erected storage tanks

SAW is predominantly an automatic welding process with mechanized equipment, which controls travel speed

Because SAW is used primarily for welding vessel seams in the flat position, positioning of seams to accommodate welding is required

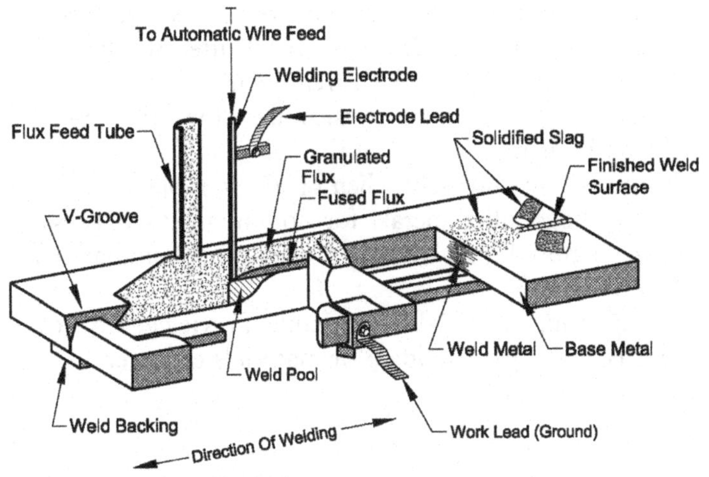

Figure 5.18 Submerged arc welding (SAW).

Requires turning rolls for welding longitudinal and circumferential shell seams and positioning for welding head seams

A wide range of deposition rates can be obtained with the process, and even higher productivity can be obtained when more than one electrode is used (e.g., tandem electrodes); welds are generally of high quality and have good mechanical properties when the proper welding consumables (wire/flux combination) are selected

Consumables

Filler Metals

In general, the main requirements for a filler metal are as follows:

Compatibility with the base material

- Mechanical properties that meet or exceed the mechanical properties of the base metal after welding (and when required, after heat treatment)
- Satisfactory corrosion resistance for the process environment
- Produces a sound weld when properly applied

Recommended filler metals for commonly used pressure vessel materials and welding processes are summarized in Figure 5.19

Other filler metals may be acceptable if reviewed on a case-by-case basis by a welding or materials engineer

Flux

Used in several different forms depending on the welding process:

- For the SMAW process, it is extruded as a coating on the electrodes
- For the FCAW process, it is added as a power inside the coded electrode
- For the SAW process, it is deposited on the work ahead of the welding arc as a loose granular material

No flux is used for the GTAW and GMAW processes

Flux serves some or all of the following functions:

- Stabilizes the welding arc
- Shields the arc and molten metal from atmosphere oxygen and nitrogen contamination
- Forms a slag blanket over the molten weld metal to help hold it in position and provide further shielding from contamination during cooling
- Provides deoxidizers (Mn and Si) to improve weld metal soundness and alloying elements (e.g., Cr, Mo, or Ni) for local alloy welds

FILLER SELECTION SUMMARY FOR COMMON PRESSURE VESSEL APPLICATIONS					
	SMAW	GTAW	GMAW	FCAW-G (note 1)	SAW
CS Vessels	E7018 per AWS A5.1 (note 2)	ER70S-2/3/6 per AWS A5.18	ER70S-2/3/6 per AWS A5.18	E7XT-1/5 per AWS A5.20 (note 3)	F7P2-EA1/EA2 (PWHT'd) or F7A2-EM12K (as welded) per AWS A5.17 (note 4)
C-½ Mo Vessels	E7018-A1 per AWS A5.5 (note 2)	ER80S-D2 per AWS A5.28 (note 5)	ER80S-D2 per AWS A5.28 (note 5)	E70T5-A1 or E8XT1-A1 per AWS A5.29 (notes 3 & 6)	F7X2-EA1/EA2 per AWS A5.23 (note 4)
1 ¼ Cr-½ Mo Vessels	E8018-B2 per AWS A5.5	ER80S-B2 per AWS A5.28	ER80S-B2 per AWS A5.28	E8XT1-B2 or E8XT5-B2 per AWS A5.29 (notes 3 & 6)	FXXX-EB2 per AWS A5.23 (note 4)
2 ¼ Cr-1 Mo Vessels	E9018-B3 per AWS B5.5	ER90S-B3 per AWS A5.28	ER90S-B3 per AWS A5.28 (note 5)	E9XT1-B3 or E9XT5-B3 per AWS A5.29 (notes 3 & 6)	FXXX-EB3 per AWS A5.23 (note 4)
5 Cr-½ Mo Vessels	E502 per AWS A5.4	ER502 per AWS A5.9	ER502 per AWS A5.9	E502T-1 per AWS A5.22 (notes 3 & 6)	F7X2-EB6 per AWS A5.23 (note 4)

Figure 5.19 Filler metal selection summary for common pressure vessel applications.

Type 304L SS Vessels	E308L per AWS A5.4	ER308L per AWS A5.9	ER308L per AWS A5.9	E308LT-1 per AWS A5.22 (note 6)	ER308L (note 7)
Type 316L SS Vessels	E316L per AWS A5.4	ER316L per AWS A5.9	ER316L per AWS A5.9	E316LT-1 per AWS A5.22 (note 6)	ER316L (note 7)
Type 347/321 SS Vessels	E347 per AWS A5.4	ER347 per AWS A5.9	ER347 per AWS A5.9	E347T-1 per AWS A5.22 (note 6)	ER347 (note 7)

Notes:

1. Filler metals shown are for the FCAW-G process only. Do not use the FCAW-SS process for pressure vessel fabrication unless reviewed on a case-by-case basis by a welding or materials engineer.

2. Only low hydrogen electrodes (E7018. etc.) are acceptable for CS and alloy steel welds. Cellulosic flux electrodes (E6010, etc.) and other non-low hydrogen electrodes are not acceptable. Cellulosic flux electrodes may, on rare occasions, be used for root passes of one-sided welds if carefully reviewed by a welding or materials engineer.

3. Only low hydrogen FCAW electrodes are acceptable for CS and alloy steel welds. "Low hydrogen" is defined as, "less than 10 ml/100 gm of diffusible hydrogen in typical as-deposited weld metal, measured by the mercury displacement or gas chromatograph methods per AWS A4.3." Note that this requirement Is supplementary to (and not covered by) applicable AWS specifications.

Figure 5.19 *(continued)*.

4. Wire and flux combinations for SAW processes should be selected for appropriate strength level and composition, since SAW flux can add significant alloy content to weld metal.
5. C-1/2Mo welds with 80 ksl filler metals must be stress relieved for wet H2S services to prevent sulfide stress cracking.
6. Use of FCAW-G processes for alloy and stainless steel welds should only be accepted after careful review by a welding or materials engineer.
7. There is no AWS specification for SAW filler metals for austenitic stainless steels. Use Lincoln ST-100 or Undo 80 flux. Other fluxes should be approved by a welding or materials engineer.

Figure 5.19 *(continued)*.

Shielding Gas

Inert shielding gases (argon and helium) are required for the GTAW process to avoid contamination of the tungsten electrode

The most common shielding gases for the GMAW and FCAW-G processes are Ar, CO_2, and combinations of the two

The exact shielding gas composition is selected to provide the desired bead shape and to minimize spatter (100% CO_2 shielding is cheaper but produces greater amounts of spatter)

Other less common shielding gases are also acceptable

Ar-O_2 mixtures are sometimes used for low-alloy steels (e.g., 2{1/4} Cr-1 Mo), and He-Ar-CO_2 mixtures are sometimes used for austenitic stainless steels (i.e., Type 304L)

Backing Gas

Inert backing gas (such as Ar, CO_2, or possibly N_2 for stainless-steel vessels) is required with GMAW and

GTAW processes for root passes of one-sided welds on alloys with more than 3% Cr content

Inert backing gas is required to prevent high-temperature oxidation of Cr (sometimes called *sugaring*) on the inner surface of the weld

To use inert backing gas for pressure vessel fabrication is usually difficult and can entail purging large volumes

Special precautions are usually necessary to localize the purging area, and one-sided welds should only be used where absolutely necessary

Preheat and Interpass Temperature Control

Preheating

Process of raising the temperature of the base metal above ambient temperature immediately before welding and holding it during welding

Not recommended for stainless steels

Preheating carbon and alloy steels accomplishes the following function:

- Reduces residual stresses, shrinkage, and distortion by minimizing differential thermal expansion between the weld metal and base metal; this is especially important for welds or thick sections or welds with high restraint
- Reduces weld metal and heat-affected zone (HAZ) hardness by slowing down the cooling rate; the driving temperature gradient between the molten weld metal and base metal, as well as the thermal conducting of the base metal, is reduced by preheat (Figure 5.20)
- Increases the diffusion rate of hydrogen from the weld metal and HAZ; this helps prevent hydrogen embrittlement cracking (also called *underbead cracking* or

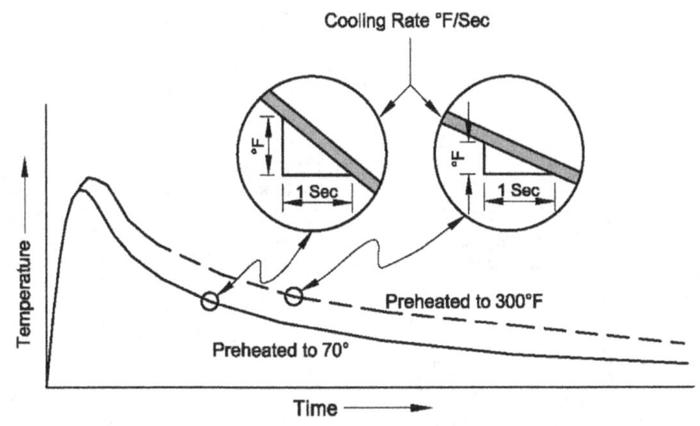

Figure 5.20 Effects of preheating.

delayed hydrogen cracking) of the weldment when it cools to ambient temperature

Minimum preheat temperatures for carbon and alloy steels are primarily determined by the following:

The hardenability of the weld metal and HAZ

This is determined by a calculated "carbon equivalent (CE)" factor; basically, the greater the CE, the greater the required preheat; see Figure 5.21 for the effect of hardenability (or CE) and preheat on underbead cracking susceptibility

Minimum preheat temperatures for carbon and alloy steels are primarily determined by the following:

The amount of restraint in the weldment; for pressure vessels, this is primarily dependent on metal thickness

The hydrogen charging characteristics of the fluxes or electrodes; this is not usually a consideration if low-hydrogen electrodes or fluxes are used; if non-low-hydrogen fluxes or electrodes are required for some reason, then the minimum required preheat should be

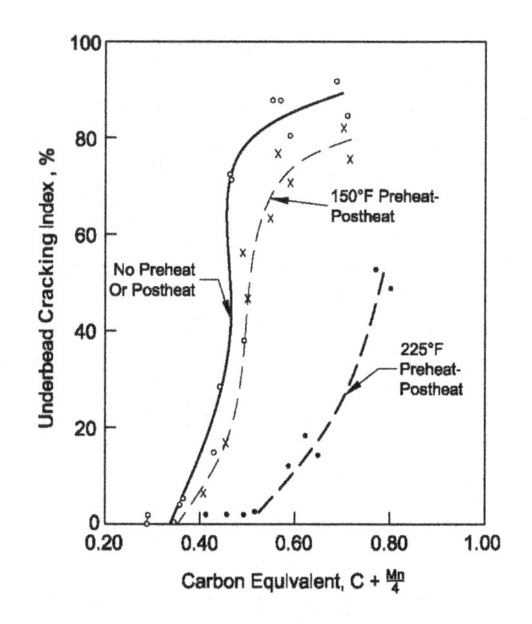

Note:

This Figure Is An Example Of The Effect Of Preheating On Underbead Cracking Susceptibility For A Weld Bead Deposited On A Test Plate.This Figure Should Not Be Used For Selecting Preheat Temperatures For Pressure Vessel Welds. Figure 600-22 Should Be Used Instead

Figure 5.21 Effect of preheat and carbon equivalent on underbead cracking susceptibility.

increased appropriately; consult a welding or materials engineer

Minimum recommended preheat temperatures are summarized in Figure 5.22 for common pressure vessel materials

Note that these recommendations exceed the ASME Code minimum values for some materials

Interpass Temperature Control

Process of controlling the temperature of the deposited weld metal between specified limits during multipass welds

	Shell Thickness, In.	Minimum Preheat Temperature °F	Interpass Temperature Limits °F
CS vessels	to 1.25 >1.25	50 200	50-450 200-500
C-1/2Mo vessels	to 0.625 > 0.625 to 2 >2	50 250* 300*	50-450 250*-550 300*-600
1 1/4Cr-1/2Mo vessels	all	300*	300*-600
2 1/4Cr-1Mo vessels	all	300*	300*-600 (note 1)
5Cr-1/2Mo vessels	all	450*	450*-650 (note 1)
Austenitic SS Vessels	all	none	300* max. (note 2)

* = Exceeds ASME Code minimum requirements

Notes:

1. For 2¼ Cr-1Mo and higher alloy steels, the minimum preheat temperature should be maintained continuously until all welding is completed, and until all welds are stress relieved. (This is to prevent delayed hydrogen cracking of nontempered material at ambient temperature.) An acceptable alternative is to perform an "intermediate stress relief" (ISR) at 1000°F to 1200°F for 30 minutes before allowing the metal to cool below the preheat temperature. The material

Figure 5.22 Preheat and interpass temperature recommendations for common pressure vessel materials.

must still be fully stress relieved at a later time. Note that many fabricators prefer a low-temperature "dehydrogenation heat treatment" (DHT) at 500°F to 600°F for several hours. This procedure should only be accepted after very careful review of welding and fabrication techniques, and only with high-quality fabrication shops.

2. No preheat should be used, and interpass temperatures should be minimized by use of stringer passes. Austenitic stainless steels (such as the 300 series SSs) can lose corrosion resistance because of excessive sensitization during slow cooling.

Figure 5.22 *(continued)*.

The lower limit is usually the same as the minimum preheat temperature and accomplishes the same functions

An upper limit is also relevant (especially when utilizing high-heat input processes) to prevent the weld metal from becoming too hot during subsequent weld passes

If the metal is allowed to get too hot during subsequent welding passes, the following detrimental effects can occur

- Excessive grain growth of the weld metal and HAZ can adversely affect the impact properties of carbon and alloy steel weldments
- Carbon and low-alloy steels can experience precipitates of nitrites (called *blue embrittlement*), which will adversely affect their impact properties
- Austenitic stainless steels (such as the 300 series) can lose corrosion resistance because of excessive sensitization during slow cooling
- Columbium stabilized austenitic SS (such as Type 347, Incoloy 825, and Alloy 20) are susceptible to solidification cracking (also called *hot-short cracking*) during slow cooling

Recommended interpass temperature limits were summarized in Figure 5.22 for common pressure vessel materials

Note that the recommended lower limits exceed the ASME Code minimum preheat temperature for some materials

Post-Weld Heat Treatment (PWHT)

Any type of thermal cycling after welding is called *PWHT*

Stress-Relief Heat Treatment

Most common form of PWHT

Involves heating the material to a temperature high enough to significantly relax residual stresses from welding, but low enough to avoid metallurgical phase transformation

Heat treatment at this temperature accomplishes the following functions

Reduces Residual Welding and Forming Stresses

This is the main benefit of any PWHT

Reducing residual stresses improves resistance to corrosion and all forms of stress corrosion cracking

It allows for dimensional stability during machining and improves mechanical properties like impact toughness and ductility

See Figure 5.23 for an example of how various PWHT temperatures and holding times affect residual stress levels in weldments; note that PWHT temperature is much more important than holding time

Softens (or "Tempers") Hard Metal HAZ

This only applies to hardenable carbon and alloy steels; austenitic stainless steels are not thermally hardenable

See Figure 5.24 for the effect of various PWHT temperatures and holding times on weld metal hardness; note that

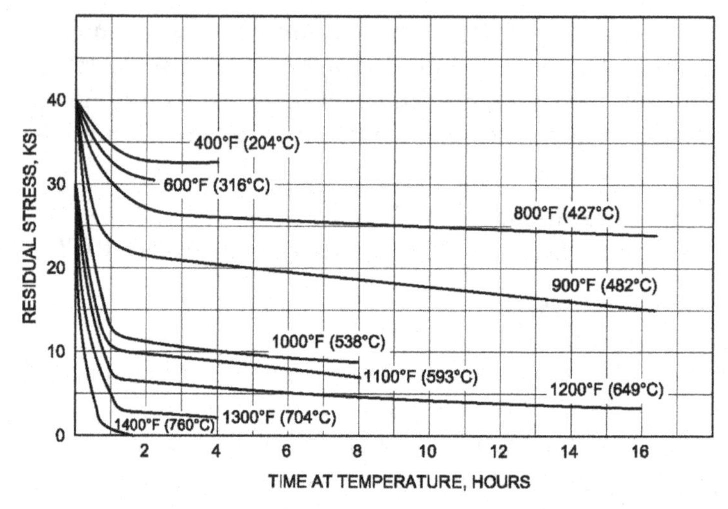

Note :

This figure is an example of the effect of PWHT temperature and time on the residual stress levels for carbon-manganese steel. This figure should not be used for selecting PWHT temperatures.

Figure 5.23 Influences of PWHT temperature and time on relieving stress for carbon-manganese steel.

PWHT temperature is much more important than holding time

Out Gasses Hydrogen from the Weld Metal

Out gassing helps prevent delayed hydrogen cracking of hardenable carbon and alloy steel weldments when cooled to ambient temperature

The need for stress relief depends on several factors; some of the applications where stress relief should be specified are summarized here:

1. All carbon steel welds greater than 1.5 inches thick should be stress relieved to reduce residual stresses and to improve impact properties (ASME Code requirement)

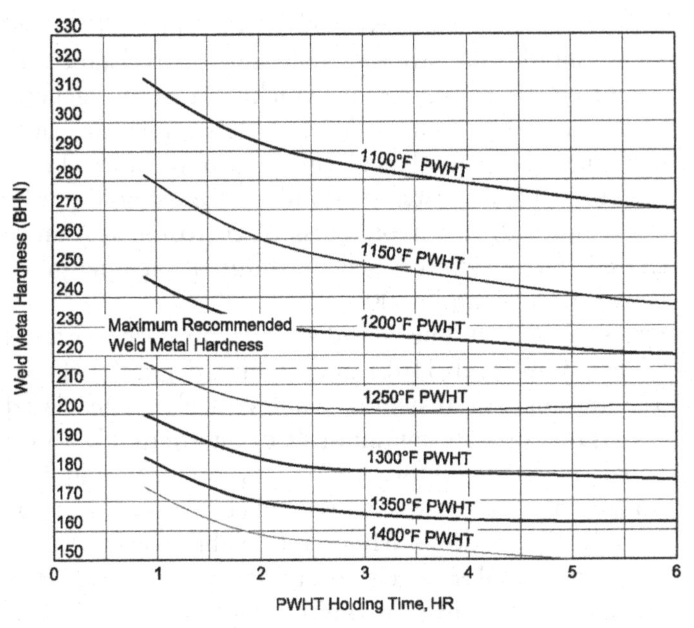

Note:

This Figure Is An Example Of The Effect Of PWHT Temperature And Time On Weld Metal hardness For 2¼ Cr-1 Mo Material. The Hardness Valves Shown Represent Average Valves. This Figure Should Not Be Used For Selecting PWHT Temperatures. Figure 600-25 Should Be Used Instead.

Figure 5.24 Effect of PWHT temperature and time on weld metal hardness for 2{1/4} CR-1mo materials.

2. All c-{1/2}Mo welds greater than {5/8} of an inch thick should be stress relieved to reduce weld metal and HAZ hardness and improve impact properties (ASME Code requirement)
3. All chrome-moly alloy steel welds should be stress relieved to reduce weld metal and HAZ hardness and improve impact properties (ASME Code requirement)

4. Many austenitic stainless-steel welds and cold-formed parts should be stressed relieved for resistance to "chloride stress corrosion cracking"; some general guidelines are presented later; consult a material engineer for addition guidance; regular carbon grades (Types 304 and 316) of austenitic stainless steels should not be stress relieved; this avoids loss of corrosion resistance caused by excessive sensitization at temperatures between 800°F and 1500°F
5. Stress relief should be considered for low-carbon grades (Types 304L and 316L) and stabilized grades (Types 321 and 347) of austenitic stainless steels for insulated vessels that operate continuously or intermittently above 150°F (even "low-chlorine" types of insulation will concentrate chlorine); stress relief should also be considered for low-carbon and stabilized grades of austenitic stainless steels for un-insulated vessels that operate continuously or intermittently above 150°F if they are in "high-chloride" environments (examples: costal location and offshore platforms with salt spray, locations with saltwater fire spray systems, and process streams with high-chlorine contents)
6. Carbon steel welds and cold-formed parts in boiler feed water de-aerators that use steam should be stress relieved to avoid "corrosion fatigue"
7. Carbon steel and austenitic stainless-steel welds and cold-formed parts should be stress relieved for resistance to "caustic stress corrosion cracking" (also called *caustic embrittlement*) if above 140°F for concentrations from 1 wt% to 30 wt% caustic, and if above 110°F for concentrations greater than 30 wt%
8. Carbon steel welds and cold-formed parts should be stress relieved in the following services:

Amines (DE, MEA, MDEA, DIPA, etc.) at any temperature, for resistance to "amine sec"
Concentrated anhydrous ammonia at any temperature, for resistance to "ammonia sec"
Potassium carbonate at any temperature, for resistance to "carbon sec"
Hydrofluoric acid at any temperature, for resistance to "hydrogen embrittlement cracking"
Sour (wet H_2S) service, for resistance to "sulfide sec" (also called *hydrogen embrittlement cracking*)

Minimum recommended stress relief temperatures and holding times are summarized in Figure 5.25

Alternatives to Stress-Relief Heat Treatment

1. Higher preheat temperatures for carbon and carbon-moly steels
2. Temper bead (also called *half-bead*) welding for carbon, carbon-molybdenum, and manganese-molybdenum steels
3. Peening
4. Vibration stress relief

These alternatives are considered to be inferior to conventional stress relief

The limitations of each alternative should be understood before accepting the alternative in lieu of conventional stress-relief heat treatment

Annealing, Normalizing, and Quenching

Other higher-temperature PWHTs are also available, but are very uncommon for pressure vessel welds

They would typically only be used when the metallurgical structure of the weld must be changed to more closely match that of the base metal.

STRESS RELIEF HEAT TREATMENT RECOMMENDATIONS FOR COMMON PRESSURE VESSEL MATERIALS			
	Shell Thickness, in.	Stress Relief Temperature Range, °F (Note 1)	Holding Time
CS vessels	≤ 2 > 2	1100-1200 1100-1200	1 hr/in, 1 hr minimum, 2 hr plus 15 min. for each additional In. over 2 in.
C-½Mo vessels	≤ 2 > 2	1175*-1250 1175*-1250	1 hr/in., 1 hr minimum 2 hr plus 15 min. for each additional in. over 2 in.
1¼ Cr-½Mo vessels	≤ 5 > 5	1300*-1375 1300*-1375	1 hr/in., 2 hr minimum 5 hr plus 15 min. for each additional in. over 5 in. (notes 2 and 3)
2¼ Cr-1Mo vessels	≤ 5 > 5	1300*-1375 1300*-1375	1 hr/in., 2 hr minimum 5 hr plus 15 min. for each additional in. over 5 in. (notes 2 and 3)
5Cr-½ Mo vessels	≤ 5 > 5	1325*-1400 1325*-1400	1 hr/in., 2 hr minimum 5 hr plus 15 min. for each additional in. over 5 in. (notes 2 and 3)
Austenitic SS vessels	All	1550-1650	1 hr/in., 1 hr minimum (note 4)
* = Exceeds ASME Code minimum requirements			

Figure 5.25 Stress-relief heat treatment recommendation for common pressure vessel materials.

Notes:

1. Longer holding times at lower temperatures, per ASME Code, are not acceptable.
2. Maximum bulk weld metal hardness should not exceed 215 BHN after stress relief.
3. For ASME Code Class 2 materials, the maximum bulk weld metal hardness can frequently be relaxed to 225 BHN and the minimum stress relief temperature to a lesser value recommended by the fabricator. (This is necessary to allow for at least two future heat treatments, while still maintaining the higher minimum strength requirements for Class 2 materials.)
4. Material should be rapidly air-cooled from PWHT temperature. Austenitic stainless steels (such as the 300 series SSs) can lose corrosion resistance because of excessive sensitization during slow cooling.

Figure 5.25 *(continued)*.

These higher-temperature heat treatments reduce the strength of the vessel materials to very low levels at the highest temperature

Most vessels are not capable of even supporting their own weight; therefore, these treatments are rarely performed on completed vessels unless special supports are used

Common for vessel plate forgings and for foremen heads and shell course ring sections prior to final vessel fabrication

The main difference between them is the cooling rate; for carbon and alloy steels, the faster the cooling rate, the harder and stronger the material

Annealing processes utilize a slow furnace cooling process to create very soft material

Normalizing processes use still air for cooling to produce material with normal "hot-formed" hardness and strength, but slightly improved impact properties

Quenching processes utilize water, oil, or air jets to rapidly cool the material and thereby yield a hard and brittle, high-strength material

Welding Procedure Qualification

General Considerations

All welding procedures used for pressure vessels should be qualified per ASME Code, Section IX

Welding procedures qualified per Section IX must include two basic documents

Welding Procedure Specifications (WPS)

Specify basic guidelines for making the weld

Procedure Qualification Record (PQR)

Proves by mechanical testing that the WPS is capable of producing acceptable weld quality, thereby "qualifying" the WPS

Document that specifies the critical welding parameters (called *essential variables* by Section IX) that are required to produce the specific type of weld covered by the WPS

Changes to the essential variables will significantly affect the characteristics of the weldment; a separate PQR is therefore required for each combination of essential variables

One WPS may be supported by more than one PQR and more than one PQR may be used to support more than one WPS

Changes to essential variables require "requalification" of the WPS, which means that an alternate PQR must be complicated

Figure 5.26 is an example of a completed WPS document

Document that proves by mechanical testing that a WPS is capable of producing acceptable weld quality, thereby "qualifying" the WPS

Welding Processes and Procedures

QW-482 (Back)
WPS No.________________ Rev.____________

POSITION (QW-406) Position(s) of Groove All Welding Progression: Up x Down Position(s) of Files	POST-WELD HEAT TREATMENT (QW-407) Temperature Range None Time Range
PREHEAT (QW-406) Preheat Temp. Min 50°F* Interpass Temp. Max Not controlled Preheat Maintenance (continuous of special heating where applicable should be recorded) *Below 50°F, preheat warn to touch (100°F min)	GAS (QW-408) Percent Composition Gas(es) (Mixture) Flow Rate Shielding None Trailing None Backing None

ELECTRICAL CHARACTERISTIC (QW-409)
Current AC or DC DC Priority Reverse
Amps (Range) See below Volts (Range) See below
(amp and volt range should be recorded for each electrode size, position, thickness, and so on; this information may be listed in a tabular form similar to that shown below)

Tungsten Electrode Size and Type N/A
(pure tungsten, 2% thoriated, etc.)
Mode of Metal Transfer of GMAW N/A
(spray arc, short circuiting arc, etc.)
Electrode Wire Feed Spread Range N/A

TECHNIQUE (QW-410)
String or Wave Based 2-G and 4-G = String; 3G = Weave (5/8" max.)
Orifice or Gas Cup
Size
Initial and Interpass Cleaning (brushing, etc.) Brushing and Slag Pick

Method of Back Gauging Carbon-Air Arc
Oscillation
Manual
Contact Tube to Work Distance N/A
Multiple or Single Pass (per side)
Multiple
Multiple or Single Electrodes Single
Travel Speed (Range) See below
Peening None
Other

		Filler Metal		Current				
Weld Layer (s)	Process	Class	Dis	Type polar	Amp. Range	Volt Range	Travel Speed Range	Other (remarks, comments, hot wire addition, technique, torch angle, etc.)
Root, filler, and backweld passes	SMAW SMAW SMAW	E7018 E7018 E7018	3/32 1/8 5/32	Reverse Reverse Reverse	70-100A 110-160A 150-220A	20-22 V 21-24 V 22-25 V	9.8-13.4 8.7-13.0 8.5-12.8	IPM (string) IPM (string) IPM (string)

Figure 5.26 Example or a welding pressure specification (WPS).

QW-482 Suggested Format For Welding Procedure Specification (WPS)
(See QW-201.1,Section IX,ASME Boiler And Pressure Vessel Code)

Company Name Chevron Corp. By: T V Smith
Welding Procedure Specification No. 17 Date 7/15/60 Supporting PQR No.(s)
Revision No. Date 10 /18 /89 (New Form And Revisions)
Welding Process (es) SMAW Type (s) Manual
(Automatic,Manual,Machine,Or Semi-Auto)

Joints (QW-402)

Joint Design Single V
Backing (Yes) (No) X
Backing Material (Type) N/A
(Refer To Both Backing And Retainers)

☐ Metal ☐ Nonfusing Metal
☐ Nonmetallic ☐ Other

Sketches,Productions Drawings,Weld Symbols Or Written Description Should Show The General Arrangement Of The Parts To Be Welded.Where Applicable The Root Spacing And The Details Of Weld Groove May Be Specified.

(At The Option Of The Mfg... Sketches May Be Attached To Illustrate Joint Design,Weld Layers And Bead Requence,e.g. For Notch Toughness Procedures,For Multiple Process Procedures, etc.

Details
75°
1/16"
1/8"

Base Metals (QW-403)

P-No 1 Group No. 1 to P.No. 1 Group No. 1
Or
Specification Type And Grade
To Specification Type And Grade
Or
Chem.Analysis And Moch.Prop.
To Chem.Analysis And Moch.Prop.
Thickness Range:
Base Metal: Groove 3 /16" Through 1 1 /4" Filter
Pipe Dia. Range: Groove All Filter
Other

Filler Metals (QW-404)

Spec. No. (SFA) 5.1		
AWS No.(Cleu) E 701 B		
F-No. 4		
A-No. 1		
Size Of Filler Metals 3 /32",1 /8",5/ 32"		
Deposited Weld Metal		
Thickness Range		
Groove 1 1 /4" Max		
Fillet		
Electrode-Flux (Class) N/A		
Flux Trade Name N/A		
Consumable Insert N/A		
Other		

Each Bees Metal-Filter Metal Combination Should Be Recorded Individually.

This form (E00006) may be obtained from the Order Dept. ASME, 346 E, 479, NY 10017

Figure 5.26 (*continued*).

A welder completes a weld on a coupon of the material specified by the WPS

All essential variables specified by the WPS must be used

Type of welding process

QW-482 (Back)

WPS No.______ Rev.______

POSITION (QW-406)
Position(s) of Groove All
Welding Progression: Up X
Down
Position(s) of Files

POSTWELD HEAT TREATMENT (QW-407)
Temperature Range None
Time Range

PREHEAT (QW-406)
Preheat Temp. Min 50°F
Interpass Temp. Max Not controlled
Preheat Maintenance
(Continuous of special heating where applicable should be recorded)
*Below 50°F, preheat warn to touch (100°F min)

GAS (QW-408)

	Percent Composition		
	Gas(es)	(Mixture)	Flow Rate
Shielding	None		
Trailing	None		
Backing	None		

ELECTRICAL CHARACTERISTIC (QW-409)
Current AC or DC DC Priority Reverse
Amps (Range) See below Volts (Range) See below
(Amps and Volts range should be recorded for each electrode size, position, and thickness, etc. This information may be listed in a tabular form similar to that shown below.)

Tungsten Electrode Size and Type N/A (Pure Tungsten, 2% Thoriated, etc.
Mode of Metal Transfer of GMAW N/A (Spray arc, short circuiting arc, etc)
Electrode Wire feed spread range N/A

TECHNIQUE (QW-410)
String or Wave Based 2-G and 4-G = string; 3G = Weave (5/8" max.)
Orifice or Gas Cup Size
Initial and interpass Cleaning (Brushing, etc.) Brushing and slag pick

Method of Back Gauging Carbon-air arc
Oscillation Manual
Contact Tube to work Distance N/A
Multiple or Single Pass (per side) Multiple
Multiple or Single Electrodes Single
Travel Speed (Range) See below
Peening None
Other

Weld Layer (s)	Process	Filler Metal Class	Filler Metal Dis	Current Type polar	Current Amp. Range	Volt Range	Travel Speed Range	Other (e.g., Remarks, Comments, Hot wire Addition, Technique, Torch Angle, Etc)
Root, filler, & backweld passes	SMAW	E7018	3/32	Reverse	70-100A	20-22 V	9.8-13.4	IPM (string)
	SMAW	E7018	1/8	Reverse	110-160A	21-24 V	8.7-13 0	IPM (string)
	SMAW	E7018	5/32	Reverse	150-220A	22-25 V	8.5-12.8	IPM (string)

Figure 5.26 *(continued)*.

Type of base metal
Type of filler metal
Preheat temperature
PWHT

Any welding parameters that are not specified by the WPS may be set at the welder's discretion

Upon completion of welding, the weldment is subjected to tension tests, guided bend tests, and fillet weld test if applicable, as required by Section IX

Refer to Code Paragraph QW-450, Section IX, for mechanical testing requirements for PQR test specimens

Impact testing of PQR test specimens may also be required if minimum impact toughness properties are specified

The welding variables that were used for the weld and mechanical test result are recorded in the PQR

Welder's name and mechanical testing company must also be recorded in the PQR

PQR must be certified by the vessel manufacturer or contractor as a true record of the welding variables and mechanical test results

Changes to certified PQRs are not permitted, except that editorial changes may be made with recertification of the PQR

A separate PQR is required for each combination of essential variables specified by the WPS

More than one PQR may therefore be required to qualify a WPS

For example, a WPS may specify a base metal thickness range of 1/16 inch to 1 inch, which is too large a range to be qualified by a single PQR; the WPS may therefore be qualified by two PQRs:

- One PQR with a 1/8-inch-thick test coupon (T), which qualifies the WPS for thickness of 1/16 to 1/4 inch (2T)
- A separate PQR with 1/2-inch-thick test coupon (T), which qualifies the WPS for thickness of 3/16 to 1 inch (2T)

Figure 5.27 is an example of a completed PQR document

Welding Processes and Procedures

QW-483 (Back)

Tensile Test (QW-150)

Specimen No.	Width	Thickness	Area	Ultimate Total Load lb.	Ultimate Unit Stress psi	Character of Failure and Location
01 Vertical	1.000	0.726	0.726	45200	62300	Shear in plate
02 Vertical	1.000	0.741	0.741	45250	61100	Shear in plate
01 Horizontal	1.000	0.716	0.716	45200	63100	Shear in plate
01 Horizontal	1.000	0.744	0.744	45000	61000	Shear in late
01 Overhead	1.000	0.725	0.725	44200	61000	Shear in plate
02 Overhead	1.000	0.715	0.715	46150	64500	Shear in plate

Guided Bend Tests (QW-160)

Type and Figure Number	Result
Side bends QW 462.2(A) horizontal	H1, H2, no defects passed
Side bends QW 462.2(A) vertical	V1, V2, no defects passed
Side bends QW 462.2(A) overhead	O1, O2, no defects passed

Toughness Tests (QW-170)

Specimen No.	Notch Location	Notch Type	Test Temp.	Impact Values	Lateral Exp. % Shear	Lateral Exp. Mills	Drop Weight Break	Drop Weight No Break

Fillet Weld Test (QW-180)

Result – Satisfactory: Yes_______ No_______ Penetration Into Parent Metal: Yes_______ No_______

Macro-Result_______

Other Test

Type of Test_______

Deposit Analysis_______

Other_______

Welder's Name R.B. Mahoney_______ Clock No._______ Stamp No. M-2

Test conducted by: Richmond Refinery_______ Laboratory Test No._______

We certify that the statements in this record are correct and that the test welds were prepared, welded, and tested in accordance with the requirements of Section IX of the ASME Code.

Manufacturer **Chevron USA, Inc.**

Date Original 7-25-1960 By H.P. Zeh

(Detail of record of tests are illustrative only and may be modified to conform to the type and number of test required by the code.)

Retyped 12-5-77 By: DAVID GREMEN

Figure 5.27 Example of a completed procedure qualification record (PQR).

Essential Variables

Code Table QW-415 of Section IX lists the essential variables for various welding processes

The main essential variables are summarized here

QW-483 Procedure Qualification Record (PQR)
(See QW-2012.Section IX,1974 ASME Boiler And Pressure Vessel Code)

Company Name Chevron USA, Inc.
Procedure Qualification Record No. 17 Date
WPS No. 17 Retyped 12-5-77 Name Change Aud
Welding Process (es) SMAW (x) Form
Types (Manual,Automatic,Semi-Auto) Manual

Joints QW-(02)

37.5"

3/4

1/16"

1/8"

3G & 4G

2G

Groove Design Used

Base Metals QW-(03)

Material Spec. A-285
Type Or Grade Gr C
P No. 1 to P No. 1
Thickness 3/4"
Diameter
Other

Postweld Heat Treatment QW-(07)

Temperature None
Time
Other

Gas QW-(08)

Type Of Gas Or Gases None
Composition Of Gas Mixture
Other

Filler Metals QW-(04)

Weld Metals Analysis Np. 1
Size Of Electrode 3/32 x 1/8" dim
Filter Metal F No. 4
SFA Specification S.1
AWS Classification E-7018
Other

Electrical Characteristics QW-(09)

Current D.C.
Polarity Reverse
Amps. 3/32" 90 amps Volt 21-23
Other 1/8" 125-135 amps

Position QW-(05)

Position Of Groove 2-G 3-G & 4-G
Weld Progression (Uphix,Downhix) Vertical Uphill
Other

Technique QW-(10)

Travel Speed Manual
Siring Or Weave Bead String 2-G 4-G Weave 3-G
Oscilation Manual
Multipass Or Single Pass (per side) Multiple & Single
Single Or Multiple Electrodes Single
Other

Preheat QW-(06)

Preheat Temp. None
Interpose Temp. Not Controlled
Other

Figure 5.27 *(continued)*.

Types of Welding Process (SMAW, GMAW, GTAW, SAW, etc.)

A change in the type of welding process requires requalification

However, a WPS that uses two different welding processes may be qualified by two separate PQRs with each using one of the processes

On the other hand, one WPS that uses only one welding process may be qualified by one PQR, which used a combination of processes that included the one used by the WPS

Only the weld metal thickness from each process, and not the total weld thickness, is applicable. Refer to Code Paragraphs QW-200.3(f) and QW-200.4 for guidelines on welding procedure qualifications using multiple processes for WPS and PQRs

Type of Base Metal (P-Number)

Various types of base metals with similar welding characteristics are grouped together under various "P-number" categories by Section IX

P-number classifications for various base metals are listed in Code Table QW-22

Some of the more common P-numbers used for pressure vessels are summarized in Figure 5.28

A change in the P-number of the base metal(s) requires requalification

Heat Input (Current Type, Polarity, Voltage Range, Current Range, Travel Speed, etc.)

A change in the type of current (AC versus DC) or polarity (straight versus reverse) and an increase in the heat input, or an increase in the volume of weld metal deposited per unit length of weld

COMMON P-NUMBERS	
Material	P-Number
Carbon Steel	1
C - 1/2 Mo	3
1 1/4 Cr - 1/2 Mo	4
2 1/4 Cr - 1 Mo	5A
5 Cr - 1/2 Mo	5B
Type 410 SS	6
Type 410S SS	7
18-8 SS (Types 304L, 316L, 321, and 347)	8

Figure 5.28 Common P-numbers.

Type of Filler Metal (A-number and F-number)

Various types of filler metals with similar welding characteristics are grouped together under various "A-number" and "F-number" categories

"A-number" refers to the chemical composition of the filler metal

"F-number" refers to the *usability* characteristics for welders (allowable positions, amount of penetration, hydrogen content, etc.) of the filler metal

Both numbers are required to adequately classify the type of filler metal

A complete listing of A-numbers and F-numbers is included in QW-442 and QW-432

A change in the A-number or F-number of the filler metal requires requalification

Post-weld heat treatment (PWHT) (yes or no)
The addition of deletion of PWHT requires requalification

Supplementary Essential Variables (Essential When Impact Testing Is Required)

If a welding procedure requires impact testing to prove that it is capable of producing welds with acceptable notch toughness, then some additional welding parameters (called *supplementary essential variables*) must be considered essential

When impact testing is not required, these variables are not considered essential

Supplementary essential variables are listed in Code Table QW-415

The main supplementary essential variables are summarized below

Base Metal Subclassification (Group Number)

Various subclassifications of base metals with similar welding characteristics are grouped together under various "group number" categories, which are subclassifications of the P-number categories; refer to Code Table QW-422

Filler Metal Subclassification (AWS Classification)

A change in the AWS SFA filler metal classification number requires requalification

Welding Position (Vertical-Uphill Only); and If Vertical-Uphill, Welding Technique (Weave Only)

A change from any position to vertical position-uphill progression welding; or if qualified for vertical-uphill welding, a change from stringer bead to weave bead technique requires reclassification

Number of Passes (Single Pass per Side Only)

A change from multipass per side to single pass per side requires requalification; also, for automatic machine welding, a change from a single electrode to multiple electrodes, or vice versa, requires requalification

Heat Input (Current Type, Polarity, Voltage Range, Current Range, Travel Speed, etc.)

A change in the type of current (AC versus DC) or polarity (straight versus reverse), an increase in the heat input, or an increase in the volume of weld metal deposited per unit length of weld requires requalification

Preheat Temperature Maximum Value

An increase in preheat temperature of 100°F or more requires requalification

Post-weld Heat Treatment (PWHT) Holding Time and Temperature

A change in the PWHT holding time and temperature range requires requalification

Nonessential Variables

Welding parameters that are not classified as essential or supplementary essential variables do not have to be specified by the WPS and are called *nonessential variables*

If they are included in the WPS, they may be changed simply by revising the WPS; refer to Code Table QW-415

Main nonessential variables are summarized as follows:

Electrode number

Method of initial or interpass cleaning (brushing, grinding, chipping, etc.)

Method of back gouging (grinding, arc-gouging, etc.)

Groove design (V-groove, U-groove, single bevel, double bevel, fillet, etc.)

Root gap

Backing rings or backing strips

Welding Procedure Review

General Considerations

The degree of welding procedure review (if any) and who performs the review should be decided depending on the criticality of the pressure vessel

A staff engineer can preview a satisfactory review for thin-walled carbon steel pressure vessels

A material or welding engineer, familiar with ASME Code and company requirements, should review thick-walled carbon steel, alloy steel, clad, and stainless-steel vessels

For pressure vessel fabrication normally submit qualified procedure to the company for review and approval before fabrication

Each welding procedure must consist of a completed WPS and one or more completed PQR, which supports the WPS; examples of completed WPS and PQR forms are included in Figures 5.26 and 5.27

When reviewing welding procedures, it is beneficial to obtain a weld map from the fabricator, if available

Weld Map

Shows which welding procedures are to be used for each weld joint

Aids judgments concerning the criticality and inspectability of weld joints made by the proposed welding procedure

For example, some less-desirable welding processes like FCAW-SS might be deemed acceptable for noncritical welds like insulation support rings, but they would not be acceptable for longitudinal or circumferential weld seams

When reviewing welding procedures, it should be clearly transmitted to the pressure vessel fabricator whether or not the welding procedures are acceptable for fabrication (either with or without revision)

Typically results in delay and extra charges from fabricators because approval, reinjection, or required revisions to welding procedures were not clearly transmitted on the initial review

Categories of Response

The following three categories should be used to avoid confusion about what is required

Acceptable as Submitted

Notifies the vessel fabricator that the proposed welding procedure is approved as written, and that fabrication may begin

Acceptable with the Following Exception(s)

Notifies the vessel fabricator that the welding procedure is approved if some revisions to nonessential variables that do not require a separate PQR (requalification) are made to the WPS

Fabrication may begin upon completion of the revisions

These revisions can be reviewed by the company inspector

Resubmittal to the engineer is not required

Unacceptable Because

Notifies the vessel fabricator that the welding procedure is not approved and that fabrication may not begin until a revised (or separate) welding procedure has been reviewed and approved

Reasons for rejecting the procedure should be clearly indicated in the transmittal to the fabricator to avoid their resubmittal of another unacceptable welding procedure

This response should be used when the revisions required to make the procedure acceptable involve changes to essential variables of supplementary essential variables (if applicable), therefore requiring a separate PQR (requalification)

This response should also be used when essential variables or supplementary essential variables (if applicable) are incorrect or missing from the completed WPS or PQR forms

Steps for Reviewing a Welding Procedure

Step 1. Verify that each PQR referenced in each WPS is included in the package of welding procedures to be reviewed. Also ensure that all pages of WPSs and PQRs are included. A good way to do this is to sort the documents and staple the supporting PQR(s) to the applicable WPS. If the supporting PQR(s) is missing from a WPS, then the WPS cannot be properly reviewed and is therefore unacceptable

Step 2. Obtain a weld map (if available) from the fabricator to determine which WPSs are to be used for which weld joints. This will help in making judgments about whether or not a certain procedure is acceptable, as the criticality and inspectability of various weld joints can differ significantly.

Step 3. Verify that all of the essential variables and supplementary essential variables (if applicable) are included in the WPS. Some nonessential variables are usually also included but do not have to be.

If essential variables or supplementary essential variables (if applicable) are missing from a WPS, then the WPS cannot be properly reviewed and is therefore unacceptable.

Step 4. Verify that all of the essential variables and supplementary essential variables (if applicable) are included in the PQR(s) and that they agree with the variables included in the WPS. Nonessential variables in the PQR(s) do not have to agree with the variables in the WPS. If essential variables or supplementary essential variables (if applicable) that are included in the supporting PQR(s) disagree with those include in the WPS, then the WPS is not qualified and is therefore unacceptable.

Step 5. Verify that all of the essential variables and supplementary essential variables (if applicable) that are included in the WPS agree with the actual requirements for the weld(s) to be performed.

This is the most important step. Frequently, the pressure vessel fabricator's welding procedures are standardized. The WPSs include all essential variables required and are properly supported by the PQRs, but the welding procedures do not agree with company specifications or the requirements on the vessel drawings.

The checklist that follows summarizes the main items that should be verified. If the answer is no to any of the questions, then the pressure vessel welds will not be performed with a qualified WPS. The WPS is therefore unacceptable.

Yes___ No___ Does the base metal (P-number) shown on the WPS agree with the material of the pressure vessel as shown on the vessel drawings? (P-numbers for commonly used base metal materials are included in Figure 5.28.)

Yes___ No___ Does the filler metal type (A-number and F-number) shown on the WPS agree with the type of filler metal that should be used for the pressure vessel weld(s)? (Recommended filler metals for common pressure vessel materials and welding processes are summarized in Figure 5.19.)

Yes___ No___ Does the shielding gas composition shown on the WPS agree with the shielding gas composition that should be used for the pressure vessel weld(s)?

Yes___ No___ For alloys with more than 3% Cr, does the WPS show the presence, flow rate, and composition of a backing gas for single-sided welds?

Yes___ No___ For GMAW or FCAW welding processes, does the transfer mode (short-circuiting transfer versus globular, spray, or pulsed transfer) that should be used for the actual pressure vessel weld(s) agree with that shown on the WPS? Note that short-circuiting transfer is acceptable only for root passes in certain applications.

Yes___ No___ Do the thicknesses of the actual pressure vessel components and welds(s) as shown on the vessel drawings fall within the thickness ranges shown by the WPS?

Yes___ No___ Is the minimum preheat temperature shown by the WPS no more than 100°F above that recommended for the vessel material? (See Figure 5.22 for recommended preheat and interpass temperature limits for common pressure vessel materials.)

Yes___ No___ If PWHT is required for the pressure vessel weld(s), does the WPS show that PWHT is required?

Yes___ No___ If PWHT is not required for the pressure vessel weld(s), does the WPS show that PWHT is not required?

Yes___ No___ Does the base metal group number shown on the WPS agree with the material of the pressure vessel as shown on the vessel drawing? (See Table QW-422 of Section IX for group number categories.)

Yes___ No___ Does the filler metal AWS classification number (E-7018, E-6010, etc.) shown on the WPS agree with that which should be used for the pressure vessel weld(s)? (Recommended filler metals for common pressure vessel materials and welding processes are summarized in Figure 5.19.)

Yes___ No___ Is the maximum preheat temperature shown by the WPS no more than 100°F below that recommended for the vessel material? (See Figure 5.22 for recommended preheat and interpass temperature limits for common pressure vessel materials)

Yes___ No___ If PWHT is required, does the PWHT temperature and holding time recommended for the pressure vessel welds fall within the range shown on the WPS? (See Figure 5.25 for recommended stress-relief heat treatment temperature ranges and holding times for common pressure materials.)

Step 6. Verify that any nonessential variables and supplementary essential variables (when impact testing is not required) that are included in the WPS agree with the requirements for the weld(s) to be preformed and that they agree with company specification(s). The fabricator should be required to revise the WPS so that it gives proper instructions for the weld to be performed.

It is also good practice to require the fabricator to add other nonessential variables and supplementary essential variables to the WPS, when impact testing is not required to ensure that important requirements in company specification(s) are adhered to.

The WPS is the instruction sheet used by the welder making the pressure vessel weld(s). If the WPS does not contain the requirements included in company specifications, these requirements may not be adhered to when the pressure vessel welds are made.

Additions or a change to nonessential variables or supplementary essential variables when impact testing is not required simply necessitates a revision to the WPS document. A separate PQR (requalification) is not necessary. The revisions can even be short handwritten notes.

Examples of nonessential variables and supplementary essential variables, which should sometimes be added to the WPS even though not required by ASME Code, Section IX, are summarized as follows:

Back gouging and back welding requirements for single-bevel groove welds (without back gouging and back welding) should only be accepted if welder access to the back of the weld is not practical

Requirement that permanent backing rings or backing strips are not allowed, when the WPS indicates use of backing material

Weld pass sequence diagram or chart for multiple pass weld(s); this requirement is especially important when multiple processes will be used for the weld(s)

Requirement on FCAW-G welding procedure(s) for carbon steel and alloy steel vessels that only electrodes with low-hydrogen content or specific brand name(s) of carbon and alloy steel electrodes are acceptable. "Low-hydrogen" means less than 10 ml of diffusible hydrogen per 100 gm of typical as-deposited weld metal per the manufacturer's specifications, as measured by the mercury displacement or gas chromatograph methods per AWS A4.3. Note that this requirement is supplementary to (not covered by) applicable AWS filler metal specifications.

Requirement on GMAW welding procedure(s) using short-circuiting metal transfer that the WPS may be used only for root passes, and only if back gouged and back welded. Short-arc GMAW processes are low-heat input procedures, which are subject to lack-of-fusion defects. The process is acceptable only for root passes if back gouged and back welded. The process may also be allowed without back welding for root passes of circumferential piping welds for fabricated nozzles and vessel internal piping, and for non-pressure-containing vessel internals.

The requirement on SMAW welding procedure(s) for carbon steel material, which indicates cellulosic (i.e., E-6010) electrodes, is that the WPS may be used only for the root pass of one-sided nonpressure containing attachment welds and circumferential piping joints and only when the base metal thickness is less than {3/4} inch. Cellulosic electrodes such as E-6010 do not have low hydrogen content. They can therefore contribute to underbead cracking of welds with high restraint, inadequate preheat, or high neat affected zone (HAZ) hardness. High restraint (joint thickness) is the main concern with carbon steel pressure vessel materials. Therefore, for joint thickness below {3/4} inch, cellulosic electrodes should be acceptable.

Step 7. Verify that the mechanical test results shown in the PQR are acceptable. Guided bend test results should say something like "satisfactory," "passed," or "acceptable." In general, this means that they had no open defects exceeding {1/8} inch after being bent. (See QW-163 of Section IX for additional details.)

The "ultimate tensile stress" for tensile test specimens should be greater than or equal to the minimum tensile strength for the base material (i.e., 70 ksi for SA-516-70 carbon steel materials, 60 ksi for SA-106 and SA-285-C carbon steel materials). Exceptions and clarifications to this ultimate tensile stress requirement are included in QW-153 of Section IX. It does not matter whether the specimens break in the weld metal or base metal.

When impact tests are required, "impact values" for the toughness of test specimens should meet or exceed the requirements of ASME Code, Figure UG-84.1, unless otherwise specified.

When the PQR is performed with a fillet-weld specimen, the "result-satisfactory" and the "penetration into parent metal" sections should be checked "yes." This means that the specimen showed no signs of cracks or lack penetration and that the sum of the visible porosity and inclusions on the fracture surface was less than {3/8} inch. The "macroresults" section should say something like "satisfactory," "passed," or "acceptable." This means that an etched cross section of the weld did not show signs of cracks or linear defects exceeding 1/32 inch and that the weld profile (concavity, convexity, etc.) was acceptable. (See QW-180 for additional details.)

Welder Performance Qualification

General Considerations

Welder performance qualification requirements should be per ASME Code, Section IX

QUALIFIED THICKNESS RANGES			
Thickness (T) of PQR Test Coupon, in.	Thickness (t) of Deposited Weld Metal of PQR Test Coupon, in.	Qualified Base Metal Thickness Range for WPS, in.	Qualified Weld Metal Thickness Range for WPS
<1/16	t	T to 2T	2t max.
1/16 to 3/8	t	1/16 to 2T	2t max.
>3/8 to <3/4	t	3/16 to 2T	2t max.
3/4 to < 1-1/2	<3/4	3/16 to 2T	2t max.
3/4 to < 1-1/2	≥3/4	3/16 to 2T	2t max.
1-1/2 to 8	<3/4	3/16 to 8	2t max.
1-1/2 to 8	≥3/4	3/16 to 8	8 In. max.
>8	t	3/16 to 1.33T	1.33t max.
Note: See Table QW-451 of ASME Code, Section IX, for limitations on qualified thickness ranges, allowable exceptions, and required mechanical test requirements.			

Figure 5.29 Qualified thickness ranges.

The purpose is to determine the ability of the welder to deposit sound weld metal

Welding operator performance qualification is used to determine the mechanical ability of the welder to operate fully automatic or machine-guided equipment

Performance Qualification Tests

Determine the ability of welders and welding operators

Each welder must be qualified for each welding process and position used in production welding

Should be performed in accordance with the variables outlined in a qualified WPS, except that preheat and PWHT requirements may be omitted

A blank welder performance qualification form is shown in Figure 5.30

Welders and welding operators are generally required to pass mechanical testing of test coupons for groove or fillet weld qualifications, but alternate rules also permit radiography for testing either groove weld test coupons or a specific length of production weld for most materials and welding processes

Mechanical tests are preferred because radiography cannot detect planar defects

All pressure vessel manufacturers or contractors are responsible for conducting the tests to qualify the performance of their welders and welding operators in accordance with a qualified WPS

The manufacturer or contractor should be responsible for the full supervision and control of the tests

Expiration

Welder qualifications have a specific time limit for when they are valid unless the welders or welding operators use the process or processes for which they are qualified

Qualification expires in 6 months if a process is not used

The fabricator is responsible for keeping records that can demonstrate that each welder is qualified by recent experience

Test Coupons

Test coupon material used for qualification does not have to be the same as designated in the WPS, and liberal substitutions are permitted

P1 material (carbon steel) can be submitted for a wide range of materials, including low-alloy steels, stainless steels, and nickel-based alloys

WELDING PROCESSES AND PROCEDURES

QW-484 SUGGESTED FORMAT FOR MANUFACTURER'S RECORD OF WELDER

OR

WELDING OPERATOR QUALIFICATION TESTS (WPQ)

(See QW-301, Section IX, ASME Boiler and Pressure Vessel Code)

Welder's Name______________________________ Clock

number__________________ Stamp no.________________

Welding process(es) used____________________________________

Type_________________________________

Identification of WPS followed by welder during welding of test

coupon____

Base material(s)

welded____Thickness________

____

Manual or semiautomatic Variables for Each Process (QW-360) Actual Values

Range Qualified

Backing (metal, weld metal, welded from both sides, flux, etc.) (QW-402)

_________ ___________

ASME P. No._____________ to ASME P. No. (QW-403)

_________ ___________

( ) Plate ( ) Pipe (enter diameter if pipe) _________

____

Figure 5.30 Welder performance qualification form.

Filler metal specification (SFA): ______________ Classification (QW-404)

__________ ____________

Filler metal F. No. __________

____

Consumable insert for GTAW or PAW

__________ ____________

Weld deposit thickness for each welding process __________

____

Welding position (1G, 8G, etc.) (QW-405) __________

____

Progression (uphill/downhill) __________

____

Backing gas for GTAW, PAW, or GMAW: fuel gas for OFW (QW-408)

__________ ____________

GMAW transfer mode (QW-409) __________

____

GTAW Welding current type/priority

__________ ____________

Machine Welding Variable for the Process Used (QW-360) Actual Values

Range Qualified

Direct/remote visual control __________

____

Figure 5.30 *(continued).*

Automatic voltage control (GTAW) __________

____

Automatic joint tracking __________

____

Welding Positions (1G, 8G, etc.) __________

____

Consumable insert __________

____

Backing (metal, weld metal, welded from both sides, flux, etc.)

__________ __________

Guided Bend Test Result

Guided Bend Tests Type ( ) QW-462.2 (side) Result ( ) QW-462.3(a) (Trans. Pl & F) Type ( )

QW-462.3(b) (Long. Fl & F) Result

Radiographic test results (QW-304 and QW-305)

____

(For alternative qualification of groove welds by radiography)

Fillet Weld – Fracture test ____________________ Length and percent of

Figure 5.30 *(continued).*

defects ______________________ in.

Macro test fusion ____________________ Fillet log else ___________in. x ___________in.

Concavity/convexity _______________in.

Welding test conducted by____

Mechanical test conducted by____

Laboratory test no. _______________

We certify that the statements in this record are correct and that the test coupons were prepared, welded, and tested in accordance with the requirements of Section IX of the ASME Code.

Organization_______________________________

Date_______________________

By___________________________

(12/48) This form (E00008) may be obtained from the Order Dept. ASME, 22 Low Drive, Box 2300, Fairfield, NJ 07007-2300

Figure 5.30 *(continued)*.

Test coupons for welder qualifications are generally either plate or pipe

Groove qualification on pipe sizes 24 inches or less qualify for pipe sizes down to 2{1/2} inches in all positions

Plate groove qualifications also qualifying pipe sizes down to 2{1/2} inches in the flat and horizontal positions

Groove weld qualifications also qualify for fillet welding but not vice versa

Test coupon thickness of less than {1/2} inch qualifies for welding on thicknesses up to two times the thickness of the test coupon, but test coupon thicknesses {1/2} inch and over are welded with a minimum of three passes because they qualify for the maximum thickness to be welded

Welder Performance Variables

Section IX categorizes welder performance variables as essential

Changes in essential variables require requalification

Not necessary to requalify welders for a new job if the essential variables in the existing welder qualification tests are not changed

Variables that are essential for performance qualification are often nonessential variables for procedure qualification and vice versa

Variables for performance qualification are listed for each welding process in Code Paragraph QW-350, Section IX, and are defined in Article IV of Section IX

Welding operating qualifications are limited by the essential variables given in Code Paragraph AW-360, Section IX, for each type of weld (fully automatic or machine-guided welding)

A summary of essential variables for common welding processes is shown in Table QW-416 of Section IX

▸ IN-SHOP INSPECTION

General Considerations

Purpose of shop inspection is to ensure equipment does the following:
- Meets specification and order requirements
- Displays good workmanship
- Free of significant damage or defects

Inspection of a fabrication plant is normally done by the following:
- Quality assurance (QA) section of purchasing
- Inspector contracted by purchasing QA to inspect vessel

Fabricator uses an "authorized inspector" to do the following:
- Verifies vessel is designed and fabricated in accordance with the ASME Code
- Sign the manufacturer's data report
- Does not check many things of interest to the company, such as the following:
  - Dimensions or orientations (except for diameter and thickness)
  - Presence of all required nozzles or internals or that they are plumb and square nozzles
  - Limitations or restrictions in company specifications on materials or welding processes that are beyond ASME Code requirements
  - Special construction details in company specifications that exceed ASME Code requirements
  - Inspection or testing requirements in company specifications that exceed ASME Code minimums

Degree of Inspection

Figure 5.31 shows degrees of inspection for pressure vessel ranging from a single inspection visit to resident inspection

Visits are listed in chronological sequence

Degrees of inspection are generally listed in accordance with the importance of the visit

Purchasing QA usually establishes the visits required based on the specifics of an order, but the responsible engineer can always participate in this decision

Figure 5.32 shows rough guidelines for choosing the appropriate dress of inspection

Figure 5.31 defines the basic purpose for each visit but does not list all of the tasks that the company inspector is required to complete during these visits

IN-SHOP INSPECTION

DEGREE OF INSPECTION

INSPECTION VISITS		1	2	3	4	5	6	7	8	9	10
PRE-INSPECTON MEETING				X	X	X	X	X	X	X	X
INTERIM AFTER MOST INDIVIDUAL LONG SEAMS ARE WELDED							X	X	X	X	X
INTERIM WHEN MOST SHELL COURSES ARE WELDED TOGETHER						X	X	X	X	X	X
INTERIM DURING NOZZLE FIT-UP AND WELDING TO VESSEL					X	X	X	X	X	X	X
INSPECT PRIOR TO PWHT (IF VESSEL IS PWHT'D)	A										
INSPECT PRIOR TO HYDRO (AFTER PWHT IF ANY)			X	X	X	X	X	X	X	X	X
WITNESS HYDROSTATIC TEST		X	X	X	X	X	X	X	X	X	X
INSPECT INTERNALLY	B										

VISIT NOTES

A This visit is required if vessel is to be PWHT'd

B This Is required:

1. For degree 1 inspection after draining, Immediately following hydrotest (not a separate visit for small vessels)
2. For vessels with internal parts that could break loose or crack welds or distort during hydro
3. For all vessels over about 1" thickness

C, D, E These visits are required if vessel is internally coated or lined. A separate pre-inspection meeting is also usually warranted for internal coatings and linings

F This visit is required for degrees 8, 9, and 10 inspection

Figure 5.31 Degrees of inspection for pressure vessel.

AFTER HYDRO, BEFORE PAINTING												if finish paint is shop applied
INSPECT INTERNAL SANDBLAST	C											
INSPECT INTERNAL PRIMER WHEN DRY	D											
INSPECT INTERNAL COATING OR LINING WHEN COMPLETE	E											
INSPECT EXTERNAL SANDBLASTED SURFACES PRIOR TO PRIMER								X	X	X	X	
INSPECT DRY PRIMER ON EXTERNAL SURFACES									X	X	X	
INSPECT DRY FINISH PAINT ON EXTERNAL SURFACES	F											
INTERIM VISIT ONCE PER WEEK TO INCLUDE ALL PRIOR ITEMS										X		

Figure 5.31 (*continued*).

RESIDENT INSPECTION											X

GENERAL NOTES

1. Add supplementary visits A through F as applicable.

2. Visits for purposes other than listed may be warranted and added (for example, to check installation of trays or other removable internals after hydrotest).

3. This matrix does nor Include all of the Inspection tasks that an inspector completes during the required plant visits shown. A list of detailed inspection tasks can be several pages long. This list is normally part of the inspection plan for a particular vessel. Inspection plans are usually prepared by Purchasing QA.

4. Each visit Is not necessarily a full 8 hour day; Inspection of a large or complex vessel prior to PWHT or prior to hydrotest may require more than 1 day.

5. Problems with quality, spec compliance, or vendor cooperation/scheduling could Increase the number of visits beyond the number shown.

Figure 5.31 *(continued)*.

Degree 4 is the recommended minimum for any vessel in hydrocarbon or critical service

Guidelines for Choosing the Appropriate Degree of Inspection for Pressure Vessels

When to Use Degree 5

1. Company specifications or standard drawings apply, and
2. Vessel
   a. is large (over 48 inches diameter or over 12 feet long), and
   b. is carbon steel, and
   c. has a moderate pressure rating (150-pound or 300-pound flanges), and
   d. is in relatively critical service

IN-SHOP INSPECTION

GUIDELINES FOR CHOOSING THE APPROPRIATE DEGREE OF INSPECTION FOR PRESSURE VESSELS

(Note that these are guidelines only)

Degree 1 when:

1. No Company specification applies, and
2. Vessel:

a. is small and simple (under 30 inches diameter), and

b. is carbon steel, and

c. has low pressure rating (under 250 psig).

Degree 2 when:

1. No Company specification applies, and
2. Vessel:

a. is of moderate size (30 inches to 48 inches diameter), and

b. is carbon steel, and

c. has low pressure rating (under 250 psig).

Degree 3 when:

1. Company specifications and/or standard drawings apply, and
2. Vessel:

a. is small and simple (under 30 inches diameter), and

b. is carbon steel, and

c. has low pressure rating (under 250 psig), and

d. relatively noncritical service (choose at least Level 4 for critical services).

Figure 5.32 Guidelines for choosing the appropriate degree of inspection for pressure vessels.

Note: "Critical Service" is difficult to define; judgment is required; factors to consider are type of service (contents of vessel), consequences of failure with regard to the process, consequences of failure with regard to personnel and surrounding facilities.

Degree 4 (this is the normal minimum for vessels in hydrocarbon service) when:

1. Company specifications and/or standard drawings apply, and
2. Vessel:

a. is of moderate size (30 Inches to 48 inches diameter), and

b. is carbon steel, and

c. has moderate pressure rating (150 pound or 300 pound flanges), and

d. is in relatively critical service.

Figure 5.32 *(continued)*.

When to Use Degree 6

1. Company specifications or standard drawings apply, and
2. Vessel
   - **a.** is large (over 48 inches diameter or over 12 feet long), and
   - **b.** is carbon steel, and
   - **c.** has shell thickness of {3/4} inch or more, and
   - **d.** has a moderate pressure rating (150-pound or 300-pound flanges), and
   - **e.** is in relatively critical service

When to Use Degree 7

Same as Level 6 but when surface preparation required for painting is “near-white” or “white metal.”

When to Use Degree 8

Same as Level 6 but when surface preparation required for painting is “near-white” or “white metal” and inspection of dry primer is judged to be warranted.

When to Use Degree 9

1. Vessel
   a. fabrication time is longer than 6 weeks, or
   b. is very large (over 10 feet in diameter or over 40 feet long), or
   c. has wall thickness over 1{1/4} inch, or
   d. has a high pressure rating (flanges 600 pounds or more), or
   e. is fabricated from low-alloy or high-alloy steels, or
   f. has complex internals, or
   g. is a trayed column

When to Use Degree 10

1. Vessel has a very heavy wall (over 4 inches thick), or
2. Vessel has highly complex internals that are inspected/tested as they are assembled and cannot be readily inspected/tested when assembly is completed.

Inspection Tasks

Figure 5.31 does not list all of the tasks that a company inspector is required to complete during the plant visits

These tasks are normally listed in an inspection plan prepared by purchasing QA

Tasks in an inspection plan can be several pages long; some of the principal tasks are as follows:

1. Verifying that welding procedures and welders are qualified per ASME Code
2. Reviewing material test reports for principal vessel parts, and verifying that vessel parts are traceable to material test reports
3. Verifying that welding procedures and preheat requirements are being followed
4. Visually inspecting all welds for flaws, contour, size, and reinforcement

5. Reading all radiographs ("X-rays") and witnessing other NDE if required
6. Providing a complete dimensional and orientation check against company drawings (when they exist and are kept up to date) or against company-reviewed vendor drawings
7. Verifying that all nozzles are present and that they are of the correct size, rating, and material
8. Verifying that internals are dimensionally correct and of the required materials
9. Reviewing the manufacturer's data report to verify that all entries are correct

Case Study 1: Construction and Inspection of Replacement Flare Knockout Drum—Most Appropriate Inspection Level: Degree 5

Background

A flare knockout drum's purpose is to remove any liquid from the gas going to the flare

If the liquid gets into the flare stack and is spewed out, it will catch fire and the burning liquid will pour down like rain into the operating unit

The engineering problem with these drums is how to design the correct size

What is the gas and liquid load that should be used for sizing, and how do you get rid of the liquid once it is in the flare knockout drum?

The liquid can be coming directly from a process, and other times it may just be accumulated liquid in the flare headers

An ideal design has all of the flare system draining downhill to the flare knockout so liquid cannot accumulate

The knockout drum dimensions are to be 6 feet (1.8 m) in diameter × 40 feet (12 m) in length

Shell and head thicknesses are set at 0.625 of an inch (16 mm)
The drum will have an internal water-sealed box
It will be constructed of mild carbon steel (SA-516-70)
The following pages are part of the "inspection plan" used for the flare knockout drum (Figure 5.33). Comments will be made on several of the required inspection points.

Preconstruction Meeting

Normally the following individuals attend these meetings:

Client (Vessel Owner)

- Client's responsible engineer
- Client's inspector(s)
- Design engineer (may be a third-party vessel designer)

Manufacturer

- Engineer
- Shop foreman
- Quality control manager

During this meeting all aspects of construction, quality, inspection, and schedule were discussed

On this particular project, because of the size of the vessel, a major concern was accurate dimensioning of large bore nozzles and external piping connections; these connections must be within tolerances so as to match existing plant piping

It was suggested that the large bore flanges be welded on from within the plant so as to adjust the alignment to existing piping; it was decided that this was not necessary

Inspector Duties

1. Verifying that welding procedures and welders are qualified per ASME Code

   After review of the Quality Control Manual, Welding Procedure Specification (WPS), and their supporting

Activity Number	Activity Description	Procedure for Performing Activity	Acceptance Standard	Notes	Cert		
						shop	client
1.0	**Pre-construction Meeting**	GIS	NA			H	H
2.0	**Engineering**						
2.1	Drawings Approved ("A" or "B")	QC Manual Section III	Client standards.			H	H
2.2	Supplier's QA/QC Procedures	QC Manual, Rev. 13	NA			H	H
3.0	**Fabrication Procedures/Qualifications**						
3.1	Welding Procedures	See List of Procedures in Notes section	ASME Section IX, QC Manual Section VII	F1-HT TS1-HT 1-HT GSA-1-HT		H	H
3.2	Welder/Operator Qualifications	QC Manual Section VII	ASME Section IX, QC Manual Section VII			H	R
3.3	Welding Repair Procedure	Weld Repair Procedure, Rev. 0	ASME Section IX			H	R
3.4	Pressure Testing Procedure	HT-2004	ASME Section VIII, UG-99,			H	H
4.0	**Nondestructive Examination Procedures**						
4.1	Radiographic Examination	GIS	ASME Section V, Article 2, UW-51,			R	R
4.2	Dry Magnetic Particle Test (MT)	MT-2004	ASME Section V ,Article 7			R	R
5.0	**Materials (Receiving Inspection)**						
5.1	Material Receiving Inspection	QC Manual Section IV(5.0)	ASME Section II,				R

Figure 5.33 Inspection plan for flare knockout drum.

5.2	Review Material Test Reports Welding consumables	QC Manual Section VII	ASME SEC. II,PART C			H	R
5.3	Review Material Test Reports – Plates, Heads, Forgings, Nozzles, Flanges, Attachments to Pressure Boundary, etc.	QC Manual Section IV	ASME SEC.II,			H	R
5.4	Material Identification – Mill Markings	QC Manual Section IV	ASME Section VIII, QC Manual			R	R
5.5	Transfer of Markings	QC Manual Section IV	ASME Section VIII, QC Manual			H	R
6.0	**Fabrication**						
6.1	Check of Plate Edges for lamination prior to welding	Visual Examination, Rev. 0	UW-31,			H	R
6.2	Welding Consumables Storage & Control	QC Manual Section VII	ASME Section II PART C			H	R
6.3	Fit-Up (Shell Plates, Heads Nozzles, etc).	Visual Examination, Rev. 0	ASME Section VIII, UW-13 & UW-33			H	R
6.4	In-Process Welding	Visual Examination, Rev. 0	Approved Welding Procedures			H	R
6.5	Welder Symbols Applied or Mapped	QC Manual Section VII	QC Manual - SECTION VII(7.1)		X	H	R
6.6	Flange/Forging Material Rating & Gasket Surface Finish	QC Manual Section V	ASME B31.3	125-250 Ra Max.		H	R
6.7	Magnetic Particle Examination all Areas Where Fit-Up Bars Were Temporarily Attachment & Arc Strikes	MT-2004	ASME Section V		X	H	R

Figure 5.33 (*continued*).

6.8	PWHT (1150 deg. F +/- 25 deg. F)	PWHT-2004	ASME Section XIII, UCS-56		X	H	H
6.9	Brinell Hardness	BH-2004		All welded seams	X	H	H
6.10	Radiographic Examination RT-1	GIS	ASME VIII UW-51		X	H	H
6.11	Final Dimensional Inspection • Orientation • Elevation • Projections • Clips/Lugs • Internal Attachments Structural Support/Base Ring Hole Pattern/Location	QC Manual Section V	Latest Revision Drawing Nos.			H	R
7.0	**Pressure Testing**						
7.1	Internal Inspection Prior to Hydrostatic Testing	Visual Examination, Rev. 0	Vessel shall be free of, slag, dirt, welding rods, and debris			H	R
7.2	Hydrostatic Testing	HT-2004	Visually Inspect All Surfaces and Welds		X	H	**H**
7.3	Dryness of Internal Surface After Hydrostatic Testing	VISUAL EXAMINATION	VESSEL SHOULD BE DRIED AFTER HYDRO TEST			H	
8.0	**Surface Preparation & Painting**						
8.1	Surface Preparation (Blast Profile)	"Sandblasting & Painting Quality Control Procedure"			X	H	R

Figure 5.33 *(continued)*.

8.2	Prime Coat	"Sandblasting & Painting Quality Control Procedure"			X	H	R
8.3	Intermediate Coat, If applicable	"Sandblasting & Painting Quality Control Procedure"			X	H	R
8.4	Final Coat, If applicable, Dry Film Thickness Measurements	"Sandblasting & Painting Quality Control Procedure"			X	H	H
8.5	Coating Records	"Sandblasting & Painting Quality Control Procedure"			X	H	H
9.0	**Manufacturer's Data – Final Dossier**					H	R
9.1	ASME Code Data Report – Complete and Signed BY AI	QC Manual Section V (7.0)	ASME Section VIII		X	H	R
9.2	Material Test Reports or Certificates of Compliance, Including Required NDT of Materials	QC Manual Section VI (5.3)	ASME Section VIII		X	H	R
9.3	NDT Reports (PT, RT, MT, UT, Hardness, Ferrite, Etc.	QC Manual Section III (4.3)	ASME Section VIII		X	H	R
9.4	Hydrostatic Test Charts & Certificate of Test	HT-2004	ASME Section VIII		X	H	R
9.5	Nameplate Facsimile (Stamping Correct)	QC Manual Section V (6.0)	ASME Section VIII		X	H	R
9.6	Inspection Release	NA	NA		X	H	**H**

Abbreviations:

I = Inspect

R/W = Random Witness

X = Documentation of inspection

W = Witness Point

R = Review

H = Hold Point

Figure 5.33 *(continued).*

Procedure Qualification Records (PQR) as well as Welder Performance Qualification (WPQ) documents, a tour of the facility was conducted to confirm that proper quality controls were being implemented and enforced

Review of Welding Documents

A simple introduction of the review of a simple WPS, PQR, and WPQ follows

The best way to begin to understand these documents is by developing a basic welding procedure step by step; the following should help to explain the important items that need to be addressed in accordance with the ASME Section IX Qualification Standard for Welding and Brazing Procedures, Welders, Brazers, and Welding and Brazing Operators

Inspection Steps

Control of Materials

1. Controlling the materials (Figures 5.34, 5.35, and 5.36) Checking the "rod oven," an important part of maintaining low-hydrogen electrodes in a dry condition, along with proper control of filler metals during welding operations so that the wrong filler metal for a welding procedure will not be issued
2. Reviewing material test reports for principal vessel parts, and verifying that vessel parts are traceable to material test reports

What Is a Mill Test Report?

A Mill Test Report (MTR) is a quality assurance document in the steelmaking industry that certifies a material's compliance with the following:

- Appropriate ASTM standards
- Applicable dimensions
- Physical and chemical specifications

Figure 5.34 Filler metal storage.

Figure 5.35 Filler metal.

Figure 5.36 Filler metal.

Why Is Material Traceability Important?

Material traceability provides "proof positive" verification, not only to the customer and the customer's contractor but to all of the governing bodies and inspection agencies that must verify that the project meets government safety requirements

When Are MTRs Needed?

Projects often require MTRs for material used in construction; some jurisdictions enforce mill certification for pressure-retaining components in boilers, pressure vessels, and pressure piping

Material traceability is also particularly important in failure investigations, to rule out product malfunction and identify components of the same material stock

Reviewing material test reports is an important function of the inspector

Not only should markings be verified, but it is also advisable to verify the chemistry for a given specification of plate, pipe, flange, fitting, or valve used during vessel construction

Figure 5.37 shows the chemistry and mechanical requirements from ASME B16.5 for an A-105 forged fitting

Now compare these table values to the MTR of a major flange manufacturer in Figure 5.38 and the flange marking in Figure 5.39

Verify the markings to ensure the proper flange is used

Welding Procedure Monitoring

3. Verifying that welding procedures and preheat requirements are being followed

These tasks can become involved—for example, one of the required WPS variables is the position of the electrical settings for current and voltage on the welding machines, and a meter is required to measure and thereby ensure that these values are within the WPS stated ranges (Figure 5.40)

When preheat is required, application of proper preheat temperature and distance from the welded joint should be confirmed; the minimum distance required will be in the Construction Code but is generally 1 inch (25 mm) in all

TABLE 1
CHEMICAL REQUIREMENTS

Element	Composition, %
Carbon	0.35 max
Manganese	0.60–1.05
Phosphorus	0.035 max
Sulfur	0.040 max
Silicon	0.10–0.35
Copper	0.40 max [Note (1)]
Nickel	0.40 max [Note (1)]
Chromium	0.30 max [Notes (1)(2)]
Molybdenum	0.12 max [Notes (1)(2)]
Vanadium	0.05 max
Columbium	0.02 max

General Note—For each reduction of 0.01% below the specified carbon maximum (0.35%), an increase of 0.06% manganese above the specified maximum (1.05%) will be permitted up to a maximum of 1.35%.

NOTES:
(1) The sum of copper, nickel, chromium and molybdenum shall not exceed 1.00%.
(2) The sum of chromium and molybdenum shall not exceed 0.32%.

TABLE 3
MECHANICAL REQUIREMENTS [Note (1)]

Tensile strength, min, psi [MPa]	70 000 [485]
Yield strength, min, psi [MPa] [Note (2)]	36 000 [250]
Elongation in 2 in. or 50 mm, min, %:	
Basic minimum elongation for walls $\frac{5}{16}$ in. [7.9 mm] and over in thickness, strip tests.	30
When standard round 2 in. or 50 mm gage length or smaller proportionally sized specimen with the gage length equal to 4D is used	22
For strip tests, a deduction for each $\frac{1}{32}$ in. [0.8 mm] decrease in wall thickness below $\frac{5}{16}$ in. [7.9 mm] from the basic minimum elongation of the percentage points of Table 4	1.50 [Note (3)]
Reduction of area, min, % [Note (4)]	30
Hardness, HB, max	187

NOTES:
(1) For small forgings, see 9.4.4.
(2) Determined by either the 0.2% offset method or the 0.5% extension-under-load method.
(3) See Table 4 for computed minimum values.
(4) For round specimens only.

Figure 5.37 Tables 1 and 3 b16.5.

Sample
Flange MTR

Sample Material Test Report For Flanges

WELDBEND

6600 South Harlem Avenue
Argo, IL 60501-1930
Phone: (708) 594-1700
Fax: (708) 458-0106

MATERIAL TEST
REPORT FLANGES

To: Customer
123 Main Street
Argo, IL
60501-1930
Attn: Joe Smith

For a glossary of terms Click Here

Description Heat I Steel Producer

Qty C Mn P S Si Cu Ni Cr Mo Cb V Ten Yield Elong RO BHN

Carbon Equivalence 0.388
Notes:
3" Class 300 Slip-on RF OXP KOPPEL Hot Formed

200 .850 .004 .013 .233 .230 .110 .060 .028 .000 .003 82500 57500 38.00 58.20 180

Figure 5.38 Material test report.

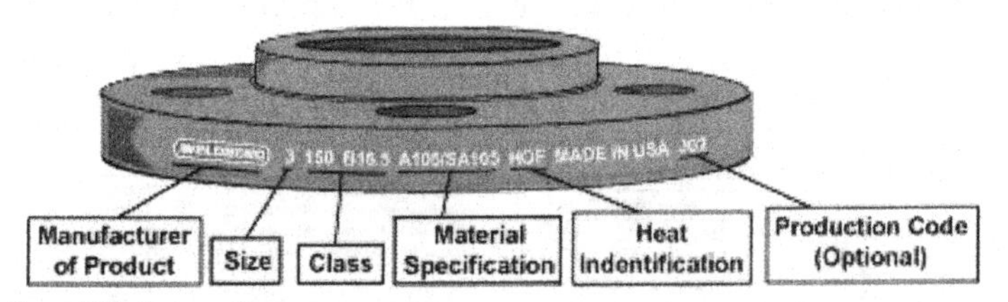

Figure 5.39 Flange markings.

directions from the weld; the WPS can require a greater distance

Common methods for measuring are crayons and thermometers

Preheat temperature indicating sticks, as shown in Figure 5.41, come in a range of melting points; simply mark the

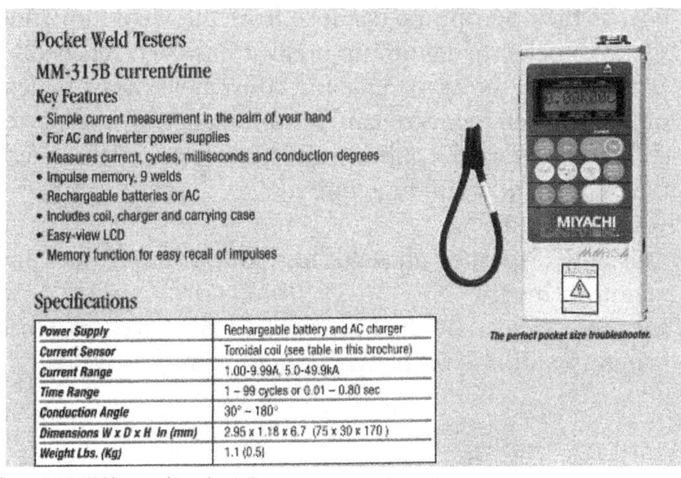

Pocket Weld Testers

MM-315B current/time

Key Features

- Simple current measurement in the palm of your hand
- For AC and Inverter power supplies
- Measures current, cycles, milliseconds and conduction degrees
- Impulse memory, 9 welds
- Rechargeable batteries or AC
- Includes coil, charger and carrying case
- Easy-view LCD
- Memory function for easy recall of impulses

Specifications

Power Supply	Rechargeable battery and AC charger
Current Sensor	Toroidal coil (see table in this brochure)
Current Range	1.00-9.99A, 5.0-49.9kA
Time Range	1 – 99 cycles or 0.01 – 0.80 sec
Conduction Angle	30° – 180°
Dimensions W x D x H In (mm)	2.95 x 1.18 x 6.7 (75 x 30 x 170)
Weight Lbs. (Kg)	1.1 (0.5)

Figure 5.40 Welding machine electrical meter.

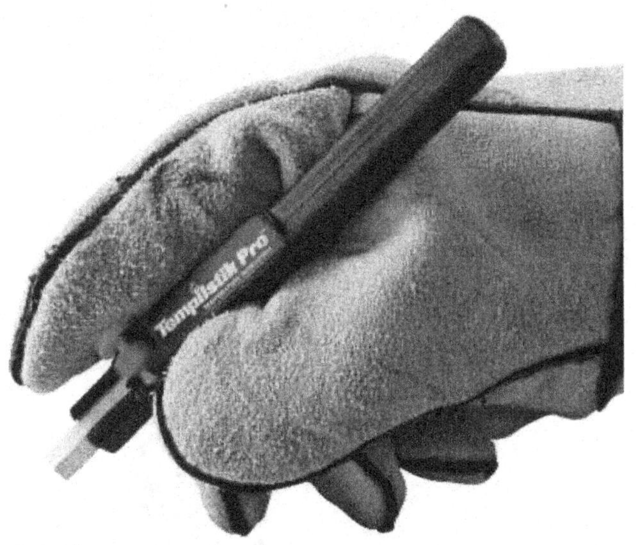

Figure 5.41 Preheat temperature stick.

part at the appropriate distance from the weld joint, and when it melts the minimum preheat has been reached

A fast, reliable, direct reading and convenient measurement of preheat temperature can be obtained with the device shown in Figure 5.42; however, like for all instrumentation, calibration will be required

Visual Inspection

4. Visually inspecting all welds for flaws, contour, size, and reinforcement

Sometimes welds should be inspected in stages as work progresses; this is particularly true for thicker welds

Figure 5.42 Infrared noncontact thermometer.

Piping and vessel codes allow weld reinforcement heights above the base metals with restrictions on height based on part thickness; what is not allowed is a finished weld that is below the base metal (concave)

Tools for the measurement of weld reinforcements are referred to as crown height gauges

Figure 5.43 shows a very well made and popular gauge

Checks Undercut, Crown Height, Porosity, Pits, All Stainless-Steel Construction

Radiographic Testing

5. Reading all radiographs ("X-rays") and witnessing other NDE if required

As an inspector, your roll may be different from that of a radiographic interpreter; this is because unless you have experience, the interpretation of radiographs can be difficult

You may see an indication, but what is it?

It is less difficult to determine if the radiographic exposure meets the requirements of the Code of Construction, such as some of the following items:

- Does the developed film meet the code density limitations imposed? (if it is too dark [black] or too light [clear], then a defect or the image quality indicator [IQI] would not be visible)
- Are location markers present in a series of radiographs to identify the exact location on a weld? (this allows any

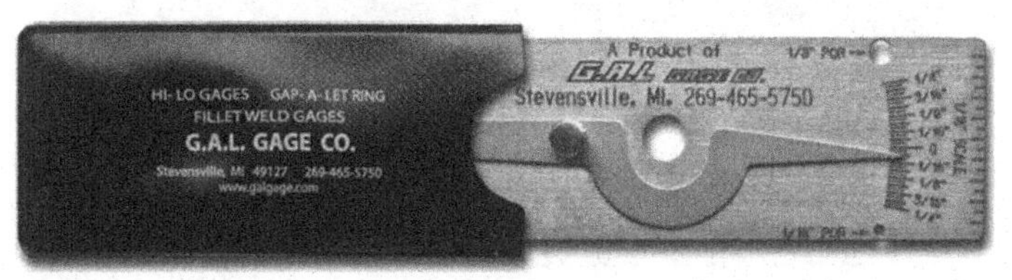

Figure 5.43 Weld and pit gauge.

defect to be found and removed, or just a comparison of the film to the weld would allow one to determine if an indication is superficial, such as grinding marks or weld concavity)

Markers A to B, B to C, and C to D appear as images on the radiographs, needed when more than one exposure is required for a weld; examples of acceptable and unacceptable exposures are shown Figure 5.44

6. Providing a complete dimensional and orientation check against company drawings (when they exist and are kept up to date) or against company-reviewed vendor drawings

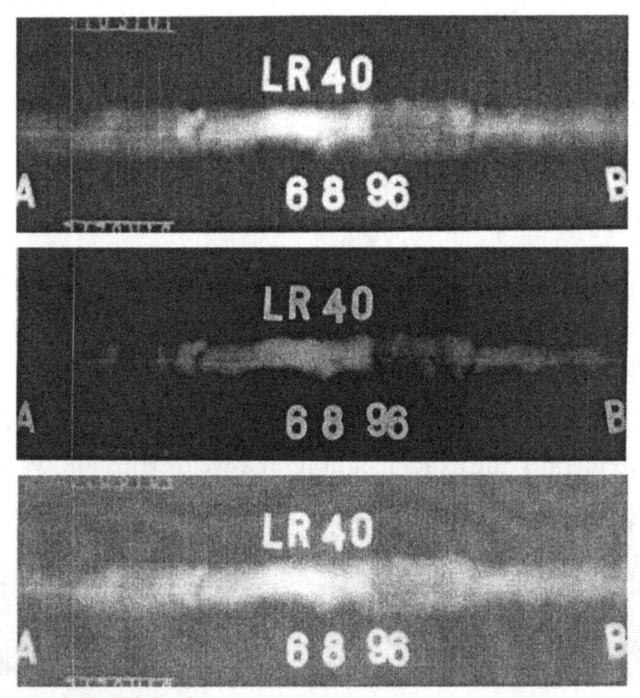

Figure 5.44 Radiographic densities.

Major concerns on a vessel's dimensions are as follows:

- Orientation of nozzles and attachments
- Elevation of nozzles and attachments
- Projection of nozzles and attachments
- Locations of internal components

After the shell courses have been rolled and when shell diameters are large, they must be maintained in a circular condition; this requires inserting a brace referred to as a spider on each end, and if the shell courses have a large diameter, then several may be spaced through the shell (shown in Figure 5.45)

Dimensional Checks

For smaller diameters and thicker walls, bracing is less critical and may only require a single brace on each end, which is used to support welding of the long seam (Figure 5.46)

After vessel heads are welded on, they will add support to the shell and the braces will no longer be necessary

Before heads are attached, a work line is marked on the circumference of the vessel shell; normally this line will be made at 12 inches (300 mm) from the chosen end of the

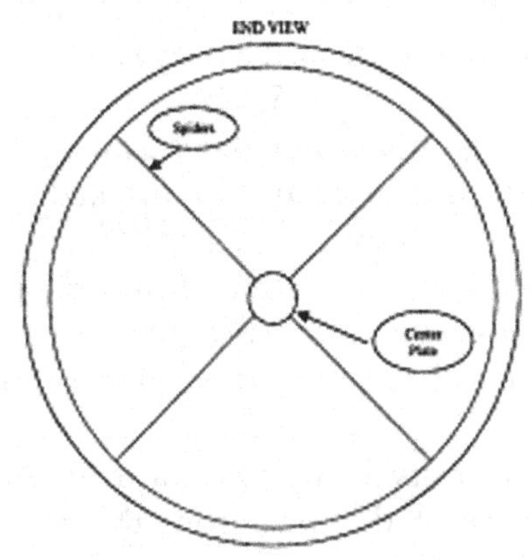

Figure 5.45 Temporary spider support.

Figure 5.46 Internal brace.

shell, and this work line is used to determine the elevation of a vessel nozzle or attachment

A reference line must be established for measurement of the elevation of nozzles and other attachments (Figure 5.47)

From this reference line, nozzle elevations are determined and later checked (Figure 5.48)

Orientation Longitudinal Reference Lines

The vessel's shell and heads are also marked with at least 0°, 90°, 180°, and 270° reference lines (Figure 5.49)

All nozzles, manways, and attachment orientations will be measured from the nearest degree reference line for cutting holes and attachments such as support locations (Figure 5.49)

For the shell, the outside diameter of the shell is computed and the length of the arcs from the reference lines is calculated for each nozzle

Figure 5.47 Transverse work line.

Figure 5.48 Checking installed nozzle elevations.

Figure 5.49 0° orientation line.

From the longitudinal and transverse lines all points on a shell can be determined

Cutting the Opening for Nozzles

The marking of all the nozzles, manholes, or attachments is best done at the same time so the intersection of these components with existing weld seams can be avoided; this is checked with respect to the orientation plan/elevation for vertical vessels and the end view/elevation for horizontal vessels

Care shall be taken to see that the correct left-hand/right-hand sides for horizontal and top/bottom ends for vertical vessels are identified; otherwise the openings made might be placed on the wrong side

On satisfactory marking of the center points of these attachments, the cutting line on the shell will be marked as needed, and these cutting lines also have to be checked before the opening is made

Consider the case of a nozzle N1 to be located at 45° from the 0° reference line with an elevation of 48 inches (Figure 5.50)

Measure the outside circumference of the shell approximately at the location where the nozzle N1 is to be installed; the outside circumference corresponding to 45° is calculated as C × 45/360; measure out from the reference centerline at 0°, and mark a parallel line at 45°

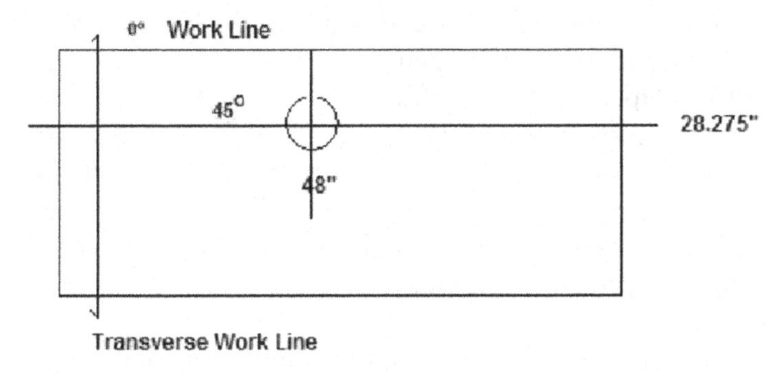

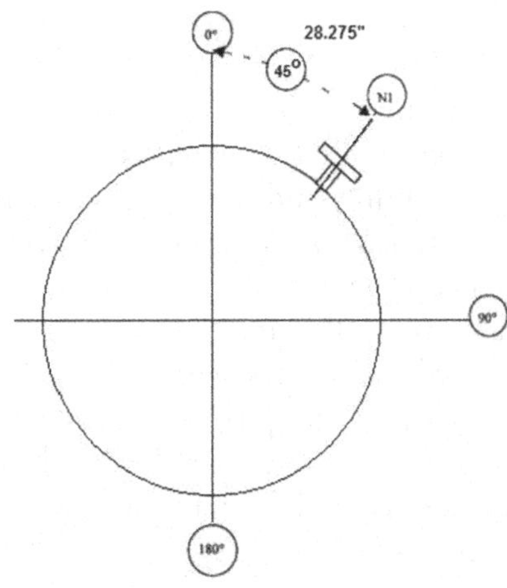

Figure 5.50 Calculate nozzle locations.

Example Nozzle Layout

The circumference of the case study vessel having a 72-inch-diameter shell is 226.195 inches; therefore, the distance to the 45° line is 226.195 × 45/360 = 226.195 × 0.125 = 28.275 inches

Measure from the 0° line to 28.275 inches around the circumference of the shell at both ends, then mark and pop a chalk line; now from the transverse work line, measure across the chalk line to the elevation of the nozzle; at this point you center punch the location of the nozzle for cutting the opening

Cutting Opening for Nozzles

Later the cutting line of the opening is also marked, taking into consideration the curvature of the shell and that of the nozzle; the in-house QC personnel shall countercheck both the center point and the cutting line to rule out possible errors in marking

Verifications

7. Verifying that all nozzles, manways, and supports are present and that they are of the correct size, rating, and material

The vessel may be near completion at the time this verification is performed; but really this should be an ongoing process as the vessel is built, and it also should be done before the vessel is coated, as markings may be difficult to read after coating is complete

Things to check include the following:

- Total number of nozzles and manways
- Nozzle material markings on pipe used to make the nozzles; any time pipe is cut, all markings should be transferred so that the original and cut piece will still be identifiable by specification

Figure 5.51 Cutting a nozzle opening (1 of 2).

- Note that often temporary lifting lugs are welded to larger heavier flanges and then must be ground off; be sure that this grinding does not remove the flange factory markings, because if this occurs after the initial confirmation, the flange markings must be restored by the manufacturer of the vessel
- Flange class rating, material, and size (Figure 5.52)

Verify the marking to ensure the proper flange is used prior to installation; this should have already been done; now we need to determine if the proper flange is on the correct nozzle

A large vessel will normally have many openings and other attachments that will require checking prior to final acceptance (Figure 5.53)

Using the vessel drawing, confirm markings; also count the total number of nozzles, openings, and supports, and verify locations; this task should have been taking place throughout the construction process because at this point in construction it will be difficult to correct

Three parties are responsible for these tasks: the shop's foreman, the manufacturer's quality control department's floor QC, and the inspector for the owner/user, amounting to a triple check of these openings and attachments

8. Verifying that internals are dimensionally correct and of the required materials

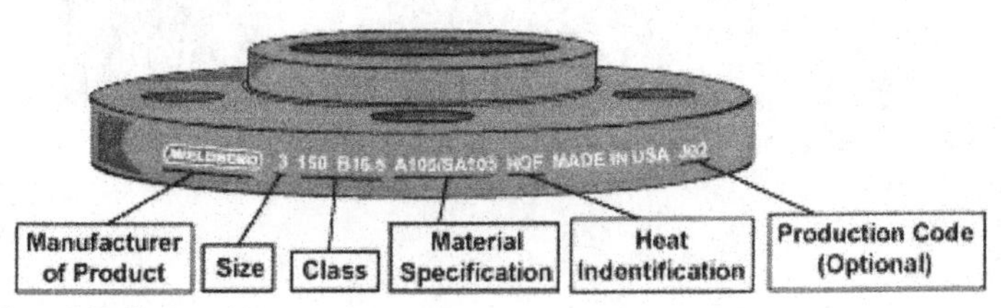

Figure 5.52 Flange markings.

Figure 5.53 Vessels have many flanged openings.

Internals may be such things as trays and down comers, de-misters, distribution manifolds, draw off manifolds, and sealed boxes

These internals will be dimensioned for proper size, verified for material, and confirmed to be properly installed at the correct location inside the vessel; in the case of a water-sealed box, it will be filled with water and tested for leakage as shown in Figure 5.54

Because venting combustibles to the flare is the drum's purpose, sometimes a water-sealed box is installed in the drum to help prevent back firing from the flare stack

This drum must be watertight (Figure 5.55), have connections for filling and draining, and include a water-level gauge glass

Figure 5.54 Water-sealed box.

Figure 5.55 Water-sealed box leak test.

This leak test was conducted for 1 hour before observing weld seams for leaks and loss of water level in the sealed box

9. Performing a leak test of reinforcement pad welds
   - Weld tests are normally conducted with air at 15 psig (approximately 1 bar)
   - In this test, a fitting is screwed into a drilled tapped hole placed in the reinforcement pad prior to welding it in place; pressure is slowly raised to 15 (1 bar), and the welds are sprayed with a bubble-forming solution on both the inside and outside of the shell and nozzle then observed for leakage
10. Ensuring that hydrostatic testing is witnessed by the authorized inspector and the inspector for the owner/user, preferably at the same time

Hydrostatic testing shall be conducted in accordance with the Code of Construction, as this vessel was constructed to the ASME Section VIII Division 1 2007 A08 Code, the test pressure was calculated as follows: 1.3 × MAWP × (stress at test/stress at design)

Stress at test and design refers to the allowable stress at two different temperatures and normally does not become significant until mild carbon steel exceeds 400°F; because our vessel material is such a carbon steel, its calculation looks as follow:

$$1.3 \times 150 \times (20{,}000/20{,}000) = 1.3 \times 150 \times 1$$
$$= 195 \text{ psig}$$

11. If the vessel is to be coated at the manufacturer's facility, confirming the film thickness of primer/top coats and any markings as required

Using magnetic pull-off film thickness gauges to determine if film thickness is in compliance with the purchaser's specification is common practice (Figure 5.56)

The inspector for the owner/user should witness film thickness measurements and determine if the coatings, primer, and top coats meet specifications as well as any required markings (Figure 5.57)

12. Reviewing the manufacturer's data report to verify that all entries are correct, confirming that the nameplate has been properly attached to the vessel, and performing the last review of the data package to accompany the vessel for items such as the following (Figures 5.58 and 5.59):
    - Material test certificate (Mill Test Reports in the case of plate) for each pressure part listed in the fabrication drawings
    - Material identification verification reports by the manufacturer QA/QC as well as appropriate third parties

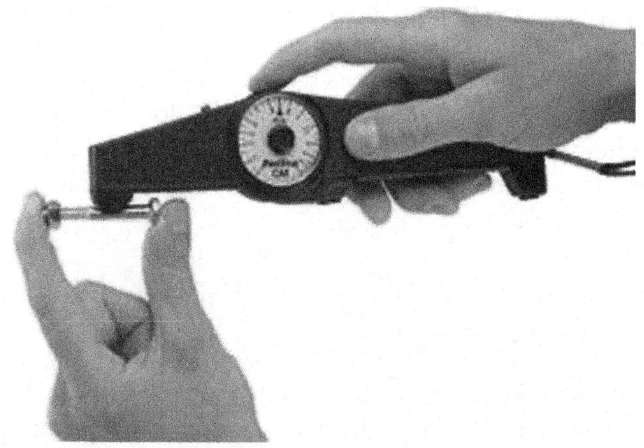

Figure 5.56 Film thickness measurement.

Figure 5.57 Marking required by client

- Incremental construction inspection reports
- Welding inspection reports, including weld maps
- Nondestructive testing (NDT) reports and a summary
- Post-weld heat treatment (PWHT) reports
- Inspection release for hydrostatic test
- Subsequent NDT reports (if applicable)
- Hydrostatic test report
- Posthydrostatic test cleaning
- Final inspection release certificate (IRC)

Summary

Verification and traceability along with inspection are the main concerns:

- Conduct a preconstruction meeting to determine the methodology that will be used for the planned vessel
- Verify the documents, welding, and all materials to be used

FORM U-1 MANUFACTURER'S DATA REPORT FOR PRESSURE VESSELS
As Required by the Provisions of the ASME Boiler and Pressure Vessel Code Rules, Section VIII, Division 1

1. Manufactured and certified by Jackson Pressure Vessels
(Name and address of Manufacturer)

2. Manufactured for Major Refining Company
(Name and address of Purchaser)

3. Location of installation Gulf Coast USA
(Name and address)

4. Type Horizontal (Horizontal, vertical, or sphere) Flare Knockout Drum (Tank, separator, jkt. vessel, heat exch., etc.) GC-1897 (Manufacturer's serial number)
None (CRN) GC-1345 Rev.C (Drawing number) NB-1302 (National Board number) 2009 (Year built)

5. ASME Code, Section VIII, Div. 1 2007 A07 (Edition and Addenda, if applicable (date)) None (Code Case number) No (Special service per UG-120(d))

Items 6–11 incl. to be completed for single wall vessels, jackets of jacketed vessels, shell of heat exchangers, or chamber of multichamber vessels.

6. Shell: (a) Number of course(s) 5 (b) Overall length 40.0 ft (12.9m)

Course(s)			Material	Thickness		Long. Joint (Cat. A)			Circum. Joint (Cat. A, B & C)			Heat Treatment	
No.	Diameter	Length	Spec./Grade or Type	Nom.	Corr.	Type	Full, Spot, None	Eff.	Type	Full, Spot, None	Eff.	Temp.	Time
1-5	72" I.D.	96"	SA-516 Gr.70	.625"	.125"	1	Full	1.0	1	Spot	.85	1150F	1 Hr.

7. Heads: (a) (Material spec. number, grade or type) (H.T. — time and temp.) (b) (Material spec. number, grade or type) (H.T. — time and temp.)

	Location (Top, Bottom, Ends)	Thickness		Radius		Elliptical Ratio	Conical Apex Angle	Hemispherical Radius	Flat Diameter	Side to Pressure		Category A		
		Min.	Corr.	Crown	Knuckle					Convex	Concave	Type	Full, Spot, None	Eff.
(a)	Ends	.625"	.125"	73.25"	4.305	2 to 1					X			
(b)														

If removable, bolts used (describe other fastening) Non-Removable
(Material spec. number, grade, size, number)

8. Type of jacket None Jacket closure None (Describe as ogee and weld, bar, etc.)

If bar, give dimensions If bolted, describe or sketch.

9. MAWP 150 psi (Internal) 15 psi (External) at max. temp. 250 F (Internal) 250 F (External) Min. design metal temp. -20F at 250 psi.

10. Impact test Exempt UCS-66 (Indicate yes or no and the component(s) impact tested) at test temperature of

11. Hydro., pneu., or comb. test pressure Hydro 195 Proof test None

Items 12 and 13 to be completed for tube sections.

12. Tubesheet None (Stationary (material spec. no.)) (Diameter (subject to press.)) (Nominal thickness) (Corr. allow.) (Attachment (welded or bolted))
(Floating (material spec. no.)) (Diameter) (Nominal thickness) (Corr. allow.) (Attachment)

13. Tubes None (Material spec. no., grade or type) (O.D.) (Nominal thickness) (Number) (Type (straight or U))

Items 14–18 incl. to be completed for inner chambers of jacketed vessels or channels of heat exchangers.

14. Shell: (a) No. of course(s) None (b) Overall length

Course(s)			Material	Thickness		Long. Joint (Cat. A)			Circum. Joint (Cat. A, B & C)			Heat Treatment	
No.	Diameter	Length	Spec./Grade or Type	Nom.	Corr.	Type	Full, Spot, None	Eff.	Type	Full, Spot, None	Eff.	Temp.	Time

15. Heads: (a) (Material spec. number, grade or type) (H.T. — time and temp.) (b) (Material spec. number, grade or type) (H.T. — time and temp.)

	Location (Top, Bottom, Ends)	Thickness		Radius		Elliptical Ratio	Conical Apex Angle	Hemispherical Radius	Flat Diameter	Side to Pressure		Category A		
		Min.	Corr.	Crown	Knuckle					Convex	Concave	Type	Full, Spot, None	Eff.
(a)														
(b)														

If removable, bolts used (describe other fastening)
(Material spec. number, grade, size, number)

Figure 5.58 Manufacturer's U-1 data report (front).

- Confirm that Code of Construction requirements are met during fabrication
- Confirm that all openings and internals have been properly placed on/in the vessel

FORM U-1 (Back)

16. MAWP N/A (Internal) ______ (External) at max. temp. ______ (Internal) ______ (External) Min. design metal temp. ______ at ______.

17. Impact test None (Indicate *yes* or *no* and the component(s) impact tested) at test temperature of ______.

18. Hydro., pneu., or comb. test pressure ______ Proof test ______

19. Nozzles, inspection, and safety valve openings:

Purpose (Inlet, Outlet, Drain, etc.)	No.	Diameter or Size	Type	Material		Nozzle Thickness		Reinforcement Material	Attachment Details		Location (Insp. Open.)
				Nozzle	Flange	Nom.	Corr.		Nozzle	Flange	
Inlet	1	30"	Full Pen	SA-106 B	A-105	0.500"	0.125	SA-516-70	UW-16.1 h		
Outlet	2	30"	Full Pen	SA-106 B	A-105	0.500"	0.125	SA-516-70	UW-16.1 h		
Instrument	3	2"	Full Pen	SA-106 B	A-105	0.365	0.125	None	UW-16.1 i		
Instrument	4	2"	Full Pen	SA-106 B	A-105	0.365	0.126	None	UW-16.1 i		
Manway	5	24"	Full Pen	SA-106 B	A-105	0.365	0.125	SA-516-70	UW-16.1 h		

20. Supports: Skirt no (Yes or no) Lugs no (Number) Legs 2 (Number) Others ______ (Describe) Attached ______ (Where and how)

21. Manufacturer's Partial Data Reports properly identified and signed by Commissioned Inspectors have been furnished for the following items of the report (list the name of part, item number, Manufacturer's name, and identifying number):

NONE

22. Remarks

Exempt from Impacts testing UG-20(f)

CERTIFICATE OF SHOP COMPLIANCE

We certify that the statements in this report are correct and that all details of design, material, construction, and workmanship of this vessel conform to the ASME BOILER AND PRESSURE VESSEL CODE, Section VIII, Division 1.

U Certificate of Authorization Number 396548 Expires 04/20/2011

Date 03/15/2009 Name Jackson Pressure Vessels (Manufacturer) Signed ______ (Representative)

CERTIFICATE OF SHOP INSPECTION

I, the undersigned, holding a valid commission issued by the National Board of Boiler and Pressure Vessel Inspectors and/or the State or Province of Texas and employed by Boiler Machinery Insurance Co. of Texas have inspected the pressure vessel described in this Manufacturer's Data Report on 03/15/2011, and state that, to the best of my knowledge and belief, the Manufacturer has constructed this pressure vessel in accordance with ASME BOILER AND PRESSURE VESSEL CODE, Section VIII, Division 1. By signing this certificate neither the Inspector nor his/her employer makes any warranty, expressed or implied, concerning the pressure vessel described in this Manufacturer's Data Report. Furthermore, neither the Inspector nor his/her employer shall be liable in any manner for any personal injury or property damage or a loss of any kind arising from or connected with this inspection.

Date 03/15/2009 Signed ______ (Authorized Inspector) Commissions NB XXXX Texas Comm YYYY (National Board (incl. endorsements), State, Province, and number)

CERTIFICATE OF FIELD ASSEMBLY COMPLIANCE

We certify that the statements in this report are correct and that the field assembly construction of all parts of this vessel conforms with the requirements of ASME BOILER AND PRESSURE VESSEL CODE, Section VIII, Division 1. U Certificate of Authorization Number N/A Expires ______.

Date ______ Name ______ (Assembler) Signed ______ (Representative)

CERTIFICATE OF FIELD ASSEMBLY INSPECTION

I, the undersigned, holding a valid commission issued by the National Board of Boiler and Pressure Vessel Inspectors and/or the State or Province of N/A and employed by ______ of ______, have compared the statements in this Manufacturer's Data Report with the described pressure vessel and state that parts referred to as data items ______, not included in the certificate of shop inspection, have been inspected by me and to the best of my knowledge and belief, the Manufacturer has constructed and assembled this pressure vessel in accordance with the ASME BOILER AND PRESSURE VESSEL CODE, Section VIII, Division 1. The described vessel was inspected and subjected to a hydrostatic test of ______. By signing this certificate neither the Inspector nor his/her employer makes any warranty, expressed or implied, concerning the pressure vessel described in this Manufacturer's Data Report. Furthermore, neither the Inspector nor his/her employer shall be liable in any manner for any personal injury or property damage or a loss of any kind arising from or connected with this inspection.

Date 3-15-2009 Signed ______ (Authorized Inspector) Commissions NB XXXX TX YYYY (National Board (incl. endorsements), State, Province, and number)

Figure 5.59 Manufacturer's U-1 data report (back).

- Verify that all required NDEs have been performed at the appropriate times and that qualified personnel were used in the examinations
- Witness final pressure testing and drying if hydrostatically tested
- Confirm coating thickness if a coating was applied

- Satisfy any other requirements that might be unique to the vessel
- Review data package for completeness
- Sign the final inspection release form if required

Case Study 2: Construction and Inspection of Replacement Flue Gas Cooler—Most Appropriate Inspection Level: Degree 9

Background

Flue gas coolers can be used in many applications; the cooler covered in this chapter is used in the flue gas of a refinery fluid catalytic cracking unit (FCCU)

The basic purposes of this cooler are to recover lost heat and use it to generate steam, much the same as a boiler

The steam generated may be sent to the utility steam header or dedicated to generating electrical power through the use of a turbo electric generator

The equipment's other purpose is to cool FCCU exhaust gases, thereby helping to control environmental emissions

Preconstruction Meeting

This particular cooler had been fabricated several times; as such it did not require the initial preconstruction meeting, as the manufacturer well understood the requirements

Material of Construction

The material of construction for this cooler's tubes was ASME P-Number 5B chrome alloy material

Typically, it is 5% chrome and 0.5% molybdenum or 9% chrome and 1% molybdenum

It was used for its suitability to long-term operation at high temperatures and its resistance to metallurgical creep

Creep

In materials science, *creep* is the tendency of a solid material to move slowly or deform permanently under the influence of stresses

It occurs as a result of long-term exposure to high levels of stress that are below the yield strength of the material; creep is more severe in materials that are subjected to heat for long periods and near melting point; creep always increases with temperature

Figure 5.60 shows the finished flue gas cooler prior to heat treatment and before the required hydrostatic test

The flue gas cooler is constructed with an inlet and outlet header for feed water input and a steam-generated outlet header (Figure 5.61)

The headers were constructed of seamless pipe using specification A335 Grade P9, a high-temperature service material (Figure 5.62)

Example of Material Confirmation

As part of the inspection task documentation such as the Mill Test Reports (MTR), the manufacturer's quality

Figure 5.60 Assembled flue gas cooler.

Figure 5.61 Flue gas inlet headers.

Figure 5.62 Identification markings.

control group performs a visual confirmation of the material marking, and the inspector for the owner/user reviews it

In jurisdictions where a construction code such as the American Society of Mechanical Engineers (ASME) is required, an authorized inspector will also confirm that the appropriate material is being used

The markings in Figure 5.62 are interpreted as follows:

- V&M: Material manufacturer's name
- Pipe Diameter: NPS 10 inches (DN250)
- Pipe Schedule: 80
- ASTM Specification: A335 Group P9 (9% chrome, 1% molybdenum)
- Heat of Steel: HT. F1305

These marking shall be confirmed by comparison to the MTR and should also be confirmed by physical testing using a positive material identification (PMI) method when the metal is an alloy

See Figure 5.63 for an example of one such method of PMI, the X-ray fluorescence technique (XRF)

Additionally thickness of the material shall be confirmed; see Figure 5.64 for an example of a digital thickness meter

XRF instruments work by exposing a sample to a beam of X-rays; the atoms of the sample absorb energy from the X-rays, become temporarily excited, and then emit secondary X-rays; each chemical element emits X-rays at a unique energy

By measuring the intensity and characteristic energy of the emitted X-rays, an XRF analyzer can provide a qualitative and quantitative analysis regarding the composition of the material being tested

Handheld XRF analyzers have the capability to quantify or qualify nearly any element from magnesium to uranium

Figure 5.63 PMI testing of materials.

Figure 5.64 A thickness gauge reading.

with handheld XRF analyzers; you have the option of taking the instrument into just about any location where the unit will fit

With this device you can find out what elements a material contains

In Figure 5.64, the digital measurement of an exchanger shell component is being performed

Ultrasonic thickness gauges can measure virtually any material such as plastics and metals (see Chapter 6 for the principle of operation)

The long span of the flue cooler tubes needed temporary supports to prevent bending or sagging while being fabricated in the horizontal position

Such support is shown in Figure 5.65; supports will be removed prior to installation of the flue gas cooler

After fabrication was completed, the flue gas cooler welds required post-weld heat treatment (PWHT)

Because of its construction, the flue gas cooler received PWHT in place, as opposed to transporting the assembly to a heat treatment oven

Figure 5.65 Temporary tube supports.

This required building an insulated structure around the inlet and outlet headers (Figures 5.66, 5.67, 5.68, and 5.69)

In Figure 5.67, the large hose is the burner's air supply from a forced draft fan; the smaller hose is the natural gas fuel supply

Each header's box assembly is equipped with its own blower (Figure 5.68)

These high-volume blowers are needed to supply the air at a rate necessary to maintain the combustion required to achieve the PWHT temperature of 1300° to 1400°F (704° to 760°C)

The gas supply carts shown in Figure 5.69 are specially designed and are used to regulate and monitor the natural gas used during the heat treatment operation

Thermocouples are attached at many locations to monitor heat treatment temperatures, and the temperatures are

Figure 5.66 Temporary heat treatment enclosure.

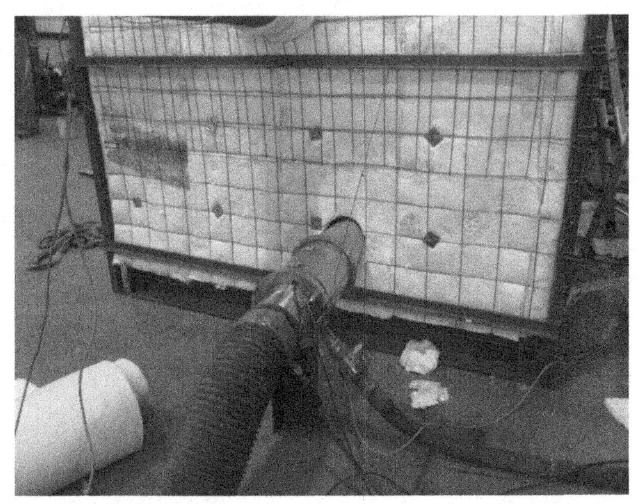

Figure 5.67 Completed burner assembly.

Figure 5.68 Forced air blower.

Figure 5.69 Gas supply cart.

recorded on a chart for documentation of heat treatment requirements

The wires shown in Figure 5.70 are the thermocouple connections going from the flue gas cooler to the chart recorder

Figure 5.71 shows the chart recorder used for this heat treatment cycle

In Figure 5.71, the chart recorder is the small box sitting on top of the larger black box

The larger black box is an electrical heat treatment power supply, not used in the flue gas cooler header heat treatment, but useful for other local heat treatments

The chart produced from the heat treatment process will be placed in the data record book for delivery to the owner/user of the flue gas cooler

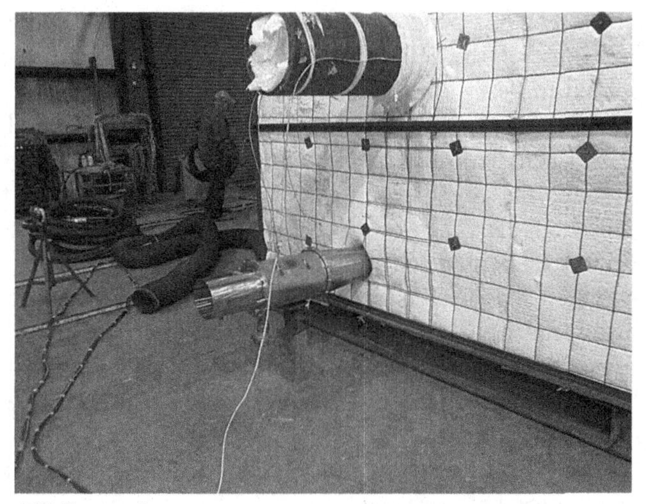

Figure 5.70 Thermocouples.

Figure 5.71 Chart recorder.

After PWHT has been completed, there is a requirement to determine if the PWHT was effective in softening the welded material on the flue gas cooler

This is confirmed by the use of hardness testing of the welds, heat affected zones, and base metal of the flue gas cooler headers; the most commonly used field method is the Brinell hardness test

Overview of Brinell Hardness Testing

The Brinell hardness lab test method consists of indenting the test material with a 10-mm-diameter hardened steel or carbide ball subjected to a load of 3000 kg

For softer materials, the load can be reduced to 1500 kg or 500 kg to avoid excessive indentation

The Brinell hardness tester used was portable hammer hit, suitable for field use

The diameter of the indentation left in the test material is measured with a handheld, low-powered microscope

The Brinell hardness number is calculated by dividing the load applied by the surface area of the indentation

Figures 5.72, 5.73, and 5.74 show some of the Brinell hardness test sites

In Figure 5.72, the indentation left by the Brinell ball will be measured using a specialized handheld, low-power scope

This measurement of the indentation diameter will be compared to a table to determine the Brinell hardness number (BHN) at this test site

The base metal test will be compared to the weld metal and weld heat-affected zone test results to assess the effectiveness of the PWHT process

In Figure 5.73, note the indentation of the Brinell ball in the center of the weld

To the left of the weld Brinell indentation, there is an indentation in the heat-affected zone

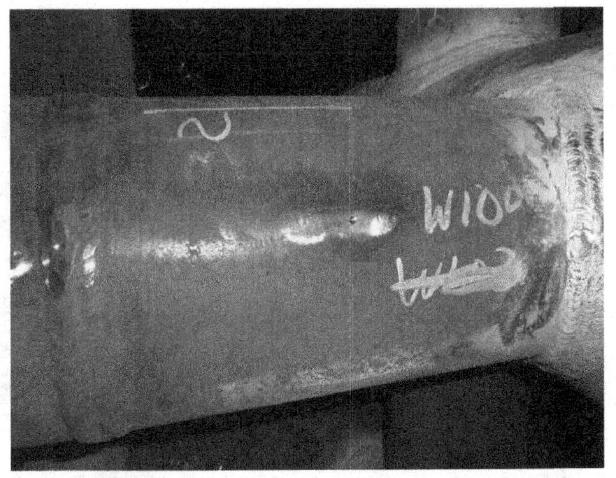

Figure 5.72 Base metal hardness testing.

Figure 5.73 Weld metal hardness testing.

Figure 5.74 Brinell evaluations.

All three areas—base metal, weld, and heat-affected zone—are measured and compared to each other and the acceptance criteria, usually given in the governing Code of Construction or in the owner/user specifications if those specifications are more stringent than those listed in the code

Figure 5.74 shows a technician using a comparator to measure the indentation of the Brinell hardness ball to determine the Brinell hardness number (BHN) at this test site

The acceptance value for this material is given in ASME Piping Code B31.3 as BHN 241; no readings exceeded the maximum allowed, and the heat treatment cycle was accepted as meeting requirements

The flue gas cooler was filled slowly and was equipped with vents in the blind flange covers (Figure 5.75) to expel the air as it was flooded with water

Figure 5.75 Hydrostatic pressure test.

However, these vents were not at the highest point in the cooler, and because the cooler was on the horizontal, not all air could be easily purged

It was not acceptable to the owner/user to weld fitting into the upper part of the donut-shaped headers for venting, and it was impractical to put the cooler on the vertical using an overhead crane because it could cause bending damage to the cooler tubes

The solution was to gently and slowly raise one end of the cooler at a time to allow as much air as possible to be expelled prior to raising the cooler to hydrostatic test pressure

Note that the gauge used is reading in its midrange, about 12 o'clock, where it is most accurate and that is was in current calibration see Figure 5.76.

Figure 5.76 Hydrostatic test pressure.

A stainless-steel cover for the finned section of the flue gas cooler was fabricated and installed after all testing had been performed and accepted

The shroud was fabricated and then split along its length to be installed in two halves over the finned section of the cooler before splitting along its longitudinal axis (Figure 5.77)

In Figure 5.78, the finned section shown is where the shroud will be placed and welded back together to complete the flue gas cooler assembly

Figure 5.79 is the flue gas cooler with the shroud welded together and installed

Figure 5.80 is the completed flue gas cooler ready for shipment to the owner's or user's facility

The flue gas cooler was dried internally after hydrostatic testing by heated forced air; after it was dry, a nitrogen

Figure 5.77 Cooler shroud before splitting.

Figure 5.78 Finned section of cooler.

Figure 5.79 Shrouds installed.

Figure 5.80 Finished cooler.

gas pad of 5 psig was placed on the internals to help prevent corrosion during storage

The flue gas cooler was shipped to the owner/user where it will be installed as a replacement for the one now in service

In-Service Inspection by Nondestructive Examination (NDE)

6

▶ OVERVIEW

This section discusses the following points:

In-service inspection of pressure vessels
Code and jurisdiction requirements
ASME Code
API Vessel Inspection Code
National Board Inspection Code
Functions of an "authorized inspector" versus those of a technician certified by the American Society of Non-Destructive Testing (ASNT)
Various forms of deterioration
In-service inspection data for each form of deterioration
Fitness-for-service analysis
Nondestructive examination (NDE) techniques
Visual examination (VT)
Dye-penetrant examination (PT)
Magnetic particle examination (MP)
Ultrasonic examination (UT)
Radiographic examination (RT)

▶ GENERAL CONSIDERATIONS

In-service inspection refers to inspection that takes place under the following conditions:
After equipment is placed *in-service*
Usually made during scheduled shutdowns

An in-service inspection program does the following:

Improves reliability and safety or operation
Increases productivity and profitability by preventing unscheduled shutdown for emergency repairs
Pressure vessels can deteriorate during normal operation and periods of process upset for several reasons:
Internal and external corrosion
Thermal aging of material
Mechanical and thermal fatigue
Stress-corrosion (environmental) cracking
Internal erosion
Hydrogen attack
Hydrogen blistering
Creep and stress rupture

Materials for Construction

Selected to minimize deterioration during service
Actual performance is not always exactly as predicted
A well-planned in-service inspection program is necessary to assure reliable and safe operation

Inspection Program

All forms of deterioration will not occur with each vessel
Established to detect only those forms of deterioration that may occur during normal operation or process upsets
Specific forms of deterioration should be determined by experienced materials engineers

Nondestructive Examination (NDE) Techniques

Can detect most forms of deterioration
Can be carried out at an early stage to permit continued operation until repair or replacement can be scheduled

▶ DESIGN FOR INSPECTION

Requirements should be considered when new equipment is designed and constructed

Weld joints should be designed with configurations and geometries that permit nondestructive examination (NDE) techniques

Selection of the most appropriate design details for weld joints requires a thorough evaluation of the following:

- Various forms of deterioration that may occur during service
- NDE procedures that can be used

Inspection during fabrication may have to be more comprehensive than required by the ASME Code so as to establish a baseline data for the in-service inspections

Difficult to distinguish between preexisting fabrication flaws from those developed during service

Requires a highly skilled NDE examiner

Lack of appropriate baseline data can complicate in-service inspection

Can increase initial vessel requirements:

- Modification or design details
- Additional inspections during fabrication
- Frequently obtained with minimal additional cost
- Additional expense

Justified by ability to assure, though in-service inspection, integrity and reliability after a long service history

▶ CODE AND JURISDICTION REQUIREMENTS

ASME Code, Section VIII Requirements

Applies directly only to the "design, fabrication and inspection during construction of pressure vessels"

Terminates when the "authorized inspector" authorizes application of the ASME Code stamp

Intent is to provide "a margin for deterioration in service so as to give a reasonably long life," and openings for visual examination or a manhole for entry are required to permit in-service inspection

Foreword makes it clear that it "deals with the care and inspection of pressure vessels only to the extent of providing suggested rules of good practice as an aid to owners and their inspectors"

In other words, the code does not directly govern the in-service inspection requirements for pressure vessels

Jurisdiction Requirements

Applies to authorities who have jurisdiction

Require owner/operators to maintain vessels in safe operating condition

Do not regulate the frequency or the types of inspection required to keep them safe

Many authorities refer to either the National Board Inspection Code or the API Pressure Vessel Inspection Code (API 510)

Most companies prefer API 510

Specifically oriented to requirements for the safe operation of vessels in the hydrocarbon processing industry

Provides greater flexibility based on the actual operating experience of the owner/operator and the expertise of that party's inspectors

API 510: Pressure Vessel Inspection Code

General Considerations

Provides useful guidance to owners/operators for meeting their obligation to maintain safe and reliable equipment

More specific about in-service inspection for internal corrosion than for any other form of deterioration that may occur
The maximum interval between internal inspections of pressure vessels is half of the remaining life related to corrosion, with a maximum of 10 years
Contains separate guidelines for pressure vessels used in oil- and gas-producing operations
In-service inspection for internal corrosion
Corrosion allowance for new vessel design
Provides design life of 20 to 30 years
Basis
Corrosion rate data obtained from lab tests
Operating experience with vessels in similar service

In-Service Inspection for Internal Corrosion

Permissible to adjust corrosion rate based on maximum corrosion rate actually exhibited by the vessel
Actual corrosion rate is the rate established at the first in-service inspection
Large vessels with two or more zones exhibiting different corrosion rates can have different inspection intervals established for each zone
Permissible to determine the depth of corrosion to satisfy these requirements while the vessel is in operation by using NDE
Vessels that are known to have a corrosion rate less than 0.005 inch/year and meet other criteria need not be inspected internally
Experience of the inspectors for the owner/operator is the primary source of information about potential forms of deterioration that can occur in a vessel under its operating conditions

National Board Inspection Code (NB Code)

Issued by the National Board of Boiler and Pressure Vessel Inspectors

Recognizes the existence of API 510 and does not claim to supersede it

Does not differ greatly from API 510 with regard to the technical requirements for in-service inspection

Less specific about forms of deterioration other than corrosion that can occur in vessels used for hydrocarbon processing

One significant difference: NB Code requires authorized inspectors employed by owners/operators to obtain a commission for the National Board and that authorities who have jurisdiction grant approval for owner/operator in-service inspection

API 510 simply permits the owner/operator to designate authorized inspectors who have appropriate qualifications and experience

NB Code also formalizes record keeping, but this does not significantly exceed normal good practice

Authorized Inspector versus ASNT-Certified NDE Technician

Authorized Inspector

Required to have knowledge of the following:

ASME Code

NB Code rules and requirements

Must determine if the condition of a vessel is satisfactory for continued service with respect to these codes

May not have the skills to satisfactorily perform NDE

ASNT-Certified NDE Technician

Must demonstrate proficiency with NDE procedures that are used to determine the extent of deterioration that has occurred in service

Cannot pass judgment on the suitability of a vessel for continued service

Both functions are usually performed by the same individual in an owner/operator organization, but this need not necessarily be the case

▶ FORMS OF DETERIORATION

General Considerations

Major forms of deterioration that can impair the integrity and reliability of pressure vessels are discussed with regard to the following:

- The process environments and service conditions that can cause the specific form of deterioration
- The physical characteristics of the specific form of deterioration that NDE can detect

Internal Corrosion

General Considerations

Most frequently encountered form of deterioration

Easiest to detect during in-service inspection, providing the following conditions are met:

- Correct locations inside the vessel are examined
- Proper NDE procedures are used

Many service environments are corrosive to the materials of construction (H_2S, CO_2, etc.)

Corrosion attack can be either of the following:

- General wastage
- Highly localized

Deposits on internal surfaces or held in crevices can trap corrosive compounds in contact with the vessel shell and can cause localized corrosion

Visual examination (VT) is usually adequate for detecting internal corrosion

Removal of the corrosion scale is required to determine the depth of localized corrosion and pitting

Ultrasonic examination (UT) can be used to detect internal corrosion and remaining thickness

Helpful to mark locations of localized corrosion or pitting on the Outside Diameter surface to guide the UT examination

Corrosion by Sulfur

Can be very corrosive to carbon steels and low-alloy steels above 500°F

Occurs as a general wastage of material, but it can be more severe at locations of high velocity or impingement

Austenitic stainless-steel cladding or weld overlays are used to protect the shell

Can cause corrosion at temperatures below the dew point of water when sulfur compounds are dissolved in a condensed aqueous phase to form an inorganic acid

Corrosion is not too severe and can be handled with a corrosion allowance or adjustments to operating conditions to prevent the condensation of water

Service environments that contain ammonia in addition to H_2S can be significantly more corrosive

Corrosion by Chlorine

Occurs only at temperatures below the dew point of water

Occurs below the liquid level unless the vapor space is below the dew point

It can be severe and highly localized

- Affects carbon steel surfaces
- Can become severely pitted

Grooves can occur where continuously moving liquid streams contact a vessel, such as overflow weirs and reflux return nozzles

Corrosion by Acids

Many crude oils contain naphthenic acid that can cause severe corrosion resembling pitting

Inorganic acids (sulfur and hydrofluoric) are used in various processes and can cause corrosion under certain process conditions

Stabilized grades of stainless steel (Types 321 and 347) can be used to minimize susceptibility to this type of corrosion

Corrosion Protection Claddings and Weld Overlays

Corrosion-resistant claddings and weld overlays

Used when a reasonable corrosion allowance is not sufficient for the corrosion rate

Protects the shell in severely corrosive environments

High-alloy materials, such as austenitic stainless steel or monel, are used

Visual examination (VT) is usually sufficient to determine the integrity of the cladding or weld overlay

Dye penetrant (PT) can supplement VT to determine if cracks are developing in the lining, and hammer testing can indicate if the lining has separated from the vessel shell

External Corrosion

Highly dependent on the natural atmospheric conditions prevailing at the geographic location

Humid seacoast locations are more corrosive than a dry inland location

Chemical emissions from a nearby plant can increase corrosivity of the atmosphere

Locations where rainwater can accumulate on vessels are especially prone to external attack

Crevices or pockets created by support, rings, and other external attachments are typical examples

Externally insulated vessels should be provided with weatherproofing to prevent rainwater from seeping into the insulation where it can be trapped against the vessel shell

Insulated vessels that operate at low temperatures between the dew point and freezing point of water may be especially vulnerable to external corrosion

Visual examination (VT) is the most appropriate procedure for detecting external corrosion

External corrosion of vessel supports, including anchor bolts, should not be overlooked

Thermal Aging

General Considerations

Deterioration by thermal aging of the material is not a serious concern

Several types of thermal aging can cause deterioration:

- Graphitization
- Temper embrittlement
- Creep embrittlement
- Sensitization of austenitic stainless steels

NDE cannot detect these types of thermal aging until cracks develop as a consequence of the deterioration

Other forms of deterioration that can occur during the long-term operation of a vessel at elevated temperatures are not included in this definition of thermal aging:

- Creep
- Hydrogen attack
- Thermal fatigue

Graphitization

Carbon steel vessels in operation at temperatures greater than 800°F are subject to graphitization

Can reduce the strength of the carbon steel plate below the minimum required by the code for the maximum allowable design stress, usually not a serious concern

Heat-affected zones of welds in the carbon steel plate can be susceptible to a much more damaging type of graphitization that can lead directly to cracking

Can be detected by NDE only after cracks have developed

Visual examination of the ID and OD surfaces can reveal these cracks before they cause failures

MT or PT can reveal the cracks at an earlier stage of development

UT can be used to detect and determine the depth of surface cracks and to detect and size internal cracks that have not yet propagated to the surface

Temper Embrittlement

Low-alloy Cr-Mo steels are susceptible to temper embrittlement after longtime operation at temperatures above 700°F

Can result in reduced C_V impact toughness at temperatures up to 250°F or higher

Brittle fracture is prevented by restricting startup and shutdown procedures to limit the pressure to 20% of the design pressure at temperatures below 250°F

Temper embrittlement cannot be detected by any NDE technique

Determined by removing "boat" samples from plates and weldments for C_V impact testing

Necessitates repair of the vessel, and should only be performed after consultation with pressure vessel and materials engineers

Creep Embrittlement

Low-alloy 1¼ Cr-½ Mo steel is susceptible to creep embrittlement when in services above 850°F

Weld heat-affected zones usually exhibit the greatest degree of embrittlement

Material becomes highly notch sensitive, and stress-rupture cracks can develop at locations of relatively high-stress concentration

Toes of nozzle welds and fillet welds for attachments appear to be the most likely locations for these cracks to occur

In-service inspection can detect cracks attributable to creep embrittlement in time to make repairs

VT of the OD and ID surfaces can detect and size cracks

MT or PT provide greater sensitivity

UT can be used to determine the depth of cracks originating at the surface and the size of cracks that originate at a stress concentration associated with an internal flaw

Sensitization of Austenitic Stainless Steel

Austenitic stainless steels in service at temperatures above 800°F can become sensitized to intergranular corrosion

Intergranular corrosion is not considered a serious problem in most service environments

Stabilized grades of austenitic stainless steel (Types 321 and 347) are used in temperatures above 750°F

PT examination of austenitic stainless-steel surfaces will usually reveal intergranular corrosion and is an especially useful testing tool for cladding and weld overlays

UT cannot detect intergranular cracks and can be difficult to use for solid stainless-steel vessels

Fine intergranular cracks are poor reflectors for ultrasonic pulses

Coarse-grain structure of austenitic stainless steels results in "noise" that complicates interpretation of ultrasonic reflections

Fatigue

Mechanical Fatigue

Caused by cyclic stresses

Is likely to occur when the vessel has experienced at least 400 stress (pressure) fluctuations that exceed 15% of the maximum allowable design stress for the material of construction

Cracks usually originate on the surface at locations of relatively high stress concentrations:

- Nozzles and the toes of fillet welds
  - Pad-reinforced nozzles and partial penetration welds are more susceptible than integrally reinforced nozzles and fill penetration welds

MT or PT can be used to detect cracks that originate on the surface of subjects to cyclic stresses

UT can be used to determine the depth of the cracks that originate at the surface

UT can also be used to detect and size the more insidious fatigue cracks that originate at internal flaws or at surfaces that are not accessible for MT or PT examination

Thermal Fatigue

Similar to mechanical fatigue, but the stresses are developed by temperature garments

Occurs at locations where the thermal garment is greatest

Almost always originates at the surface, and closely spaced cracks can develop

Nozzle reinforcement pads and partial penetration welds are especially prone to thermal fatigue

Surfaces are alternately wetter by the cooler liquids and drier by hotter vapor and nozzles where fluids are introduced at fluctuating rates at either a higher or a lower temperature than the bulk contents
MT and PT are superior for detecting these cracks
UT can be used to determine the depth of thermal fatigue cracks or to detect and size of those that might originate at a surface that is not accessible for MT or PT examination

Corrosion Fatigue

Nucleation of fatigue cracks at the stress concentrations associated with corrosion pits or the acceleration of fatigue crack growth by the simultaneous occurrence of corrosion in the crack
De-areator vessels can be highly susceptible to corrosion fatigue related to oxygen contamination of the boiler feed water, especially at low pH levels
Cracks occur in welds and heat-affected zones of vessels that have not received post-weld heat treatment (PWHT)
Detected by MT examination of the ID surface of the vessel
Surface should be lightly ground to assure adequate sensitivity for cracks detection
UT examination from the OD surface can be used to determine the depth of the cracks detected

Stress-Corrosion (Environmental) Cracking

Can result in the catastrophic failure of a pressure vessel
Dependent on the following:
Vessel material
Process environment
Operating conditions

Proper materials selection and fabrication procedures prevent stress-corrosion cracking, but the possibility of its occurrence in severe process environments should not be ignored

Major types of stress-corrosion cracking are as follows:

- H_2S stress cracking of carbon steel and low-alloy Cr-Mo steels
- Chlorine stress-corrosion cracking of austenitic stainless steels
- Ammonia stress-corrosion cracking of carbon steels
- Caustic embrittlement of carbon steel

Stress-corrosion (environmental) cracking

Cracks most often originate at the surface exposed to the process environment

Welds and weld-affected zones are susceptible

Can originate in H_2S environment below the surface

VT is not reliable as the cracks are very fine

MT or PT methods are reliable

Will reveal a great number of closely spaced cracks

Specialized UT techniques can be used to determine the depth of the cracks

Internal Erosion

Can occur when high-velocity process streams come in contact with the vessel shell

Entrainment of solid particles in the fluid stream and direct impingement of the stream on the shell increase the severity

Wear plates are used in vessel locations where high-process streams enter and impinge on the shell

Plates are usually fillet welded to the vessel

Made from a high-alloy material

VT is adequate to detect internal corrosion of a pressure vessel shell

UT examination can be used to determine the remaining shell thickness in eroded areas or to monitor erosion from the outside surface

Hydrogen Attack

Can occur in carbon steel and low-alloy Cr-Mo steels at elevated temperatures in services that contain high partial pressure of hydrogen

Reduces the strength of the material and causes micro-fissures that form until failure occurs

Weld heat-affected zones can be especially susceptible, and the attack can be more severe at locations of high stress, such as nozzles

Materials of construction are normally selected by using the Nelson curves

In-service inspection programs should assure hydrogen attack does not occur in vessels that operate at high temperatures and high pressure

Highly specialized UT techniques can be used to detect a hydrogen attack before microfissures grow to a size that can cause failure

Removing "boat" samples for analysis can reveal the hydrogen attack at an early stage, but doing so requires repair of the vessel and should only be done after consulting with materials engineer

Hydrogen Blistering

Can occur in service environments that cause hydrogen to diffuse through and "charge" the shell material with hydrogen

Process streams that contain acids or high partial pressures of hydrogen at high temperatures are likely to cause blistering

Hydrogen pressure builds up in microscopic traps, causing them to propagate and link together to form blisters in the major axis of the stringers (lies in the plane of the rolled plate)

VT examination can be used to detect blisters in the shells

Flashlight beam directed along the surface of the shell can aid in observing small blisters

UT examination can be used to determine the remaining sound shell thickness if cracks have developed in the blisters

Care should be exercised when making a UT examination to distinguish between possible lamination and inclusions at midwall of the plate and actual blisters

Creep and Stress-Rupture Cracks

Creep is the continuous plastic deformation of a material under a constant stress and at a high temperature:
700°F for carbon steel
850°F for low-alloy Cr-Mo steels
900°F for austenitic stainless steels

Stress-rupture cracks in pressure vessels operating in the creep range usually develop first at locations of high stress, such as nozzles and weldments (weld metals with heat-affected zones)

VT can be used on the surface of the vessel before failure occurs

Stress-rupture cracks propagate relatively slowly and thus can be detected at an early stage in time for repair

UT can be used to detect and size stress-rupture cracks

UT examinations should concentrate on areas of relatively high stress, such as the following:
- Nozzle welds
- Head-to-shell welds
- Longitudinal seam welds

Any hot spots that develop should receive intensive examination

ANALYSIS OF IN-SERVICE INSPECTION DATA

General Considerations

Information obtained from the inspection of a vessel must be interpreted and evaluated to determine the following:
- Vessel has sufficient integrity for continued service
- Repairs or replacement are required
- Detected flaws do not have to meet the ASME Code acceptance standards for the vessel to be suitable for continuous service without repair

ASME Code acceptance criteria
- Assure owners/users that vessels receiving the code symbol have been manufactured to a high level of quality
- Have no direct relationship to the requirements for safe operation

Some flaws occurring in service, though larger than those acceptable to ASME Code when the vessel was manufactured, are too small to cause vessel failure at design temperature and pressure

Internal Corrosion

General Considerations

Can reduce the thickness of the shell below the minimum thickness required by the ASME Code

Specified erosion allowance may not always be adequate for the actual corrosion rates that occur during operation

General Wastage

Reduces the thickness of the shell below the minimum thickness required by the code and thus can result in failure by rupture

Primary membrane stress in the corroded area will be increased above the maximum allowable design stress permitted by the ASME Code, and failure will occur when the increased stress exceeds the tensile strength of the material

Weld-joint efficiency factor used for the design should be used when calculating the minimum thickness required for the vessel shell when weld joints are within the corroded area

API 510 permits "averaging" the remaining wall thickness in corroded areas

If the average thickness is less than the minimum thickness required by the code or can be expected to be reduced below this thickness by the prevailing corrosion rate before the next scheduled inspection, either of the following should take place:

- Shell must be restored to the minimum required thickness plus a corrosion allowance by weld build-up
- Vessel must be replaced

An acceptable alternative is to re-rate the vessel

Re-rating reduces the maximum allowable working pressure so the maximum allowable design stress will not be exceeded in the corroded areas with reduced wall thickness

Localized Corrosion

Areas of localized corrosion or pitting can be permitted to have a remaining wall thickness less than the code minimum, within certain strictly defined limits

Local primary membrane stresses can exceed the maximum allowable design stress for the material of construction without significant risk of rupture, because of the "reinforcement" provided by the surrounding material

Weld joint efficiency factors used for design need not be applied when calculating minimum thickness because of the localized corrosion or pits that are farther from the edge of any weld

ASME employs joint efficiency factors to compensate for flaws that may exist in welds that do not receive full radiographic examination (RT)

Flaws in the weld metal do not reduce the strength of the plate and forging materials used for construction of the vessel

Wind and earthquake loadings need not be included in the minimum required thickness calculations because highly localized reductions in shell thickness would not be expected to affect the structural integrity of the vessel in the event of high wind or earthquake

API 510 Limitations

Limit total area of localized corrosion or pits within any 8-inch-diameter circle to a 7-inch^2 maximum

The remaining thickness of the vessel should not be less than half of the minimum required thickness at any point, and the sum of pit dimensions along any straight line within the circle cannot exceed 2 inches

ASME Code, Section VIII, Division 2, Appendix 4

Applicable to both Divisions 1 and 2 vessels

Local primary membranes are permitted to reach 1.5 times the maximum allowable design of the material used for construction

The area that has a local primary membrane stress that exceeds the maximum allowable design stress may not exceed $(Rt)^{0.5}$; and individual areas of local high stress may not be closer to one another than $2.5(Rt)^{0.5}$

These criteria allow a larger area of location corrosion than would be permitted by API 510

Areas of local corrosion or pitting that do not meet the requirements of APT 510 or ASME Appendix 4 must be repaired by weld build-up or by removal of the corroded area and insertion of a butt patch

Claddings and Weld Overlays

Visual indications that deteriorating claddings or weld overlays are resulting in corrosion of the pressure vessel shell must be further investigated

Hammer testing provides a useful indication of the integrity of the lining

"Solid" or "resonant" sound indicates that the lining has not separated from the vessel shell and that cracks have not developed

"Hollow" or "tinning" sound indicates that the liner has disbonded or that cracks have developed

Bulging or wrinkling of the lining may also occur

Deterioration of the liner does not necessarily indicate that the vessel shell is corroding and needs to be repaired, unless either of the following is true:

Continued deterioration may lead to corrosion of the shell before the next scheduled inspection,

Integrity of an internal attachment may be compromised

Corrosion of the vessel shell can be determined by removal of a section of the deteriorated lining and either of the following:

Visually examining (VT) the shell

Performing an ultrasonic examination (UT) to determine the remaining wall thickness

External Corrosion

Should be evaluated in the same manner as evaluating internal corrosion

Severe corrosion of vessel supports and anchor bolts may require evaluation by a structural engineer, especially if the vessel can be subjected to high wind or earthquake loads

Thermal Aging

Deterioration caused by thermal aging of the shell material is not usually repairable, as the thermal aging process is not reversible

Manifestations of thermal aging (usually cracking) can usually be repaired as an interim remedy to permit continued operation until the vessel can be replaced

Alternatively, it may be possible to analyze the acceptability of the crack for continued operating until a replacement vessel is obtained

Fatigue

Mechanical and thermal fatigue cracks can be repaired by first removing the cracks by grinding followed by weld metal build-up of the ground area so as to restore it to the minimum required thickness

Ground area should receive an MT or PT examination to make certain that all of the fatigue crack has been removed

The remaining shell thickness in the ground area can be determined by UT from the opposite surface using a longitudinal wave pressure

Weld build-up of the ground area may not be necessary if local primary membrane stress caused by the reduced shell thickness meets criteria in Appendix 4 of ASME Code, and Section VIII, Division 2

Alternatively, the depth of the fatigue can be determined by UT using a shear wave procedure, and a fitness-for-service analysis can be made to determine if the crack jeopardizes the integrity of the vessel

Whenever a fatigue crack is detected, the cause of the fatigue must be determined and eliminated by changes in design or operation; otherwise, it is likely that fatigue cracks will reappear in the same location after the repairs have been made

Stress-Corrosion (Environmental) Cracking

Stress-corrosion cracks in a pressure vessel can be repaired in a manner similar to fatigue cracks

Repairs can be expensive because large numbers of cracks can occur in susceptible vessels

Depth of the cracks should be determined by UT before repairs begin

Cracks could be superficial, and thus immediate repairs may not be necessary if the vessel satisfies fitness-for-service criteria

Essential to reexamine, during the next scheduled inspection, any stress-corrosion cracks that are allowed to remain in a vessel

A materials engineer should be consulted to determine the maximum advisable interval before the next examination

Internal Erosion

Should be evaluated in the same manner as localized corrosion

Wear plates should be installed whenever internal erosion is detected or existing ones should be repaired or replaced with a more protective design

Hydrogen Attack

Not acceptable to repair cracks attributable to the terminal stage of hydrogen attack by local grinding and weld build-up

Cracks are likely to develop rapidly adjacent to the repair

Fitness-for-service analysis is not applicable to cracks attributable to hydrogen attack

The component of the vessel, or the entire vessel, must be replaced before the vessel returns to service

For a limited duration it is possible to keep a vessel in service that has exhibited the initial stages of hydrogen attack detected by specialized UT examination

Reexaminations are required

Consult materials engineers

All vessels or components of vessels that are replaced due to hydrogen attack should be redesigned with a material that is more resistant to hydrogen attack

Hydrogen Blisters

May not significantly degrade the integrity of a vessel

The separation causing the blister to appear is parallel to primary membrane stress and does not reduce the load-bearing strength of the shell

Drilling a small vent hole into the blister will relieve the hydrogen pressure that builds up inside the blister and prevent it from swelling so much that cracks develop and propagate toward the surface

Only if cracks have already propagated to the surface will it be necessary to evaluate the blister further

Cracked blisters effectively reduce the thickness of the vessel shell

Remaining sound shell thickness can be determined by UT and evaluated for repair, replacement, or re-rating for a lower pressure

Creep and Stress Rupture

Can be repaired by grinding for removal and then restoring the ground area to the minimum required thickness with weld build-up, similar to that for fatigue cracks

This method may not always serve as a permanent repair, because the shell material adjacent to the repair might have undergone sufficient creep for additional stress rupture cracks to develop after a relatively short period of additional operation

May be more realistic to plan for replacement of the components of the vessel exhibiting the cracks

Stress-rupture cracks tend to propagate slowly, and if the cracks are relatively small when they are first detected, it may be acceptable to return the vessel to service until the replacement materials and components are obtained

Fitness-for-service analysis can be used to help determine if the vessel has sufficient integrity and reliability for limited continued service

FITNESS-FOR-SERVICE ANALYSIS

Introduction

Many forms of deterioration that occur in vessels are characterized by the formation of cracks

Analytical methods are available to evaluate the effect of these cracks on the integrity of the vessel if it is returned to service

Methods are commonly referred to as fitness-for-service analyses because they can predict the effect that flaws will have on the performance of a vessel under actual service conditions

Employing fitness-for-service analyses can have significant benefits:

1. Avoids the expense of unnecessary repairs
2. Allows the vessel to be returned to service until repairs or replacement can be scheduled without interrupting production
3. Plans in-service inspection programs that will prevent the failure of vessels during operation resulting in unscheduled shutdowns

An experienced pressure vessel engineer should be consulted for making a fitness-for-service analysis; this engineer should have the following qualifications:

1. Appropriate experience concerning the design, operation, inspection, and repair of pressure vessels
2. Familiarity with the effects that the process environment can have on the behavior of the materials of construction
3. Knowledge of fracture mechanics

Background

Corrosion

General or localized corrosion can reduce the wall thickness below the minimum required by the code

Both API 510 and the National Board recognize that areas of localized corrosion do not necessarily compromise the integrity of a pressure vessel, and they provide procedures for evaluating the acceptability of a local area with a thickness below the minimum required by the code

ASME Code, Section VIII, Division 2, Appendix 4

Generally accepted as a means to evaluate the effects of local corrosion

Stresses at the design pressure are allowed to reach 1.5 times the allowable design stress for the design temperature in a local area

This is, in effect, a "fitness-for-service" analysis that permits the continued use of pressure vessels that would not meet ASME Code requirements for new construction

Applies to local areas too deeply corroded to meet the reduced minimum thickness requirement or generally corroded areas that are too large to meet the criteria for a local deviation from code thickness requirement

Are repaired by weld build-up, or replaced

Indications of Cracks

Greater concern than corrosion

Normally occur at weld joints and may propagate during operation

Cracks that exceed the acceptance limits for radiographic examination in the ASME Code do not necessarily compromise the integrity of a pressure vessel

Flaws that are larger than the acceptance limits in the ASME Code will not always result in failure during start-up, nor will they necessarily propagate so readily at the operating pressure and temperature that they cause failure before the next planned shutdown

Fracture mechanics

- Technology that can predict if a crack will do one of the following:
  - Propagate as a fast fracture at the applied stress resulting from the internal operating pressure
  - Grow in a fatigue more when subjected to repeated pressure or temperature cycles

Fracture Mechanics

Based on the principles of fracture mechanics that relate crack propagation to the fracture toughness of the material, crack size, and applied stress

The following major concepts are involved :

1. Fracture will occur when critical stress intensity is exceeded at a crack tip. The critical stress intensity depends on the toughness of the material but is independent of the crack size and applied stress

2. Slow growth of a crack can occur below the critical stress intensity by a fatigue more under a cyclic stress; the rate of crack growth by fatigue depends on the cyclic change in stress intensity at the flaw tip, net on crack size, or applied stress
3. The stress intensity at the tip of a crack increases with the size of the crack at a constant, nominally applied stress but is not dependent on the toughness of fatigue properties of the material

Fitness-for-Service Analysis of Cracks or Other Flaws

Concepts of fracture mechanics are applied to determine if a vessel containing a crack larger than permitted by the code is "fit-for-service" at its intended operating condition

A critical crack size, required for failure of the vessel, can be determined for the material of construction at the vessel's operating pressure

Cracks smaller than this critical size will not cause a pressure vessel to fail when it is returned to service

Crack could grow by fatigue during operation if the vessel is subject to pressure or temperature fluctuations; the fatigue crack growth rate for the stress intensity at the crack tip must be calculated to determine if the crack will grow to a size critical enough to cause failure of the vessel prior to the next scheduled shutdown when another inspection can be made

The toughness of carbon and low-alloy pressure vessel steel depends on the temperature

Separate analyses may have to made for ambient temperature startup and elevated temperature operating conditions

Crack sizes that are acceptable for operation at elevated temperatures may present a risk of failure during startup at ambient temperature

Hydrotest could present a much more serious risk of failure than startup at ambient temperature

Although the toughness of the material at ambient temperature is the same for both hydrotest and startup, the applied stress associated with the hydrotest pressure is appreciably higher than that developed at the operating pressure

Therefore, a crack size that is entirely acceptable for both startup and operation could cause a pressure vessel to fail during hydrotest

Fitness-for-service analysis of cracks or other flaws

A hydrotest is not proof that a fitness-for-service analysis has been performed satisfactorily and should not be conducted with a vessel that contains a crack, unless a separate analysis has been made for the hydrotest conditions

Neither API 510 nor the National Board Inspectors Code provides for the fitness-for-service analysis of a pressure vessel containing a crack, but they do not explicitly prohibit making such an analysis

Information Required for Fitness-for-Service Analysis

Essential Information

Accurate sizing data for cracks (or other flaws)

Toughness and fatigue data for the material of construction at ambient and operating temperatures

Applied stress developed by the operating pressure at the location of the crack or other flaw

Cracks or other flaws can be accurately sized by UT, but the proper procedures must be used by experienced technicians

Shear wave UT is used for sizing

Time-based techniques produce more accurate results than amplitude-based techniques

Ultrasonic imaging techniques are available that can provide a permanent visual record of the size and orientation of the crack, and they should be used to support the fitness-for-service analysis of critical equipment

The toughness of the material of construction can be estimated from C_V impact data, but more accurate toughness data can be obtained directly from K_{IC} or J-integral tests and should be used whenever they are available

The applied stress developed by the operating pressure at the location of a crack or flaw must be accurately determined; the primary membrane and bending stresses used for design are usually sufficient for pressure vessel components with simple geometries; a finite-element stress analysis is required for complex geometries

Procedures for Fitness-for-Service Analysis

Scope

A fitness-for-service analysis can be made for any pressure vessel containing one or more flaws or cracks regardless of size, thickness, shape, or design conditions

Outline of the Procedure

Reference should be made to ASME Code, Section IV, Appendix A

An outline for making a fitness-for-service analysis is given next:

A. Determine flaw size and orientation
   1. Determine actual flaw size, configuration, and orientation using ultrasonic examination procedures
   2. Resolve the flaw into a simple shape by completely circumscribing with an ellipse or circle; circumscribe multiple flaws with the same ellipse or circle if the distance separating them is less
   3. Project circumscribed flaw onto a plane perpendicular to the direction of the maximum principal stress
      a. Flaws closer to a surface than 0.4 times their maximum dimension in a direction perpendicular to the surface are classified as surface flaws
      b. Flaws farther from a surface than 0.4 times their maximum dimension in a direction perpendicular to the surface are classified as internal flaws
B. Determine applied stress at the location of the flaw; applied stress at the location of the flaw should be resolved into the membrane bending components; all forms of loading (internal pressure, discontinuity, and thermal) should be considered as follows:
   1. Calculate maximum principal membrane stress at the operating pressure for cylindrical and spherical shapes using ASME Code design equations
   2. Calculate maximum principal membrane and bending stresses at nozzle opening (WRC Bulletins 107 and 297)
      a. Consideration must be given to both internal pressure and significant external piping loads; piping stress analysis may be necessary to determine piping loads
   3. Calculate principal membrane and bending stress at internal or external attachments and vessel supports (WRC Bulletin 107)

4. Calculate discontinuity stresses at head-to-shell joints and transitions resulting from internal pressure using ASME Code, Section VIII, Division 2, Appendix 4, and combine with principal membrane and bending stresses
5. Calculate thermal stresses at locations with significant temperature gradients, using ASME Code and combine with principal membrane and bending stresses
6. Use finite element stress analysis to calculate all membrane and bending stresses.

C. Determine stress intensity at flaw
   1. Using the principal membrane and bending stresses calculated earlier, calculate the stress intensity at the flaw (K,) using the following equation:

$$K_I = \sigma_m M_m(\pi)0.5\left(\frac{a}{Q}\right)^{0.5} + \sigma_b M_b(\pi)^{0.5}\left(\frac{a}{Q}\right)^{0.5}$$

Where

$\sigma_{m'}\sigma_b$ = Membrane and bending stresses, psi, in accordance with B above

A = Minor half-diameter, in., of embedded flaw; flaw depth for surface flaw

Q = Flaw shape parameter as determined from Figure 6.1 using (m + b)/ys and the flaw geometry

M_m = Correction factor for membrane stress (see Figure 6.2 for subsurface flaws, Figure 6.3 for surface flaws)

M_b = Correction factor for bending stress (see Figure 6.4 for subsurface flaws, Figure 6.5 for surface flaws)

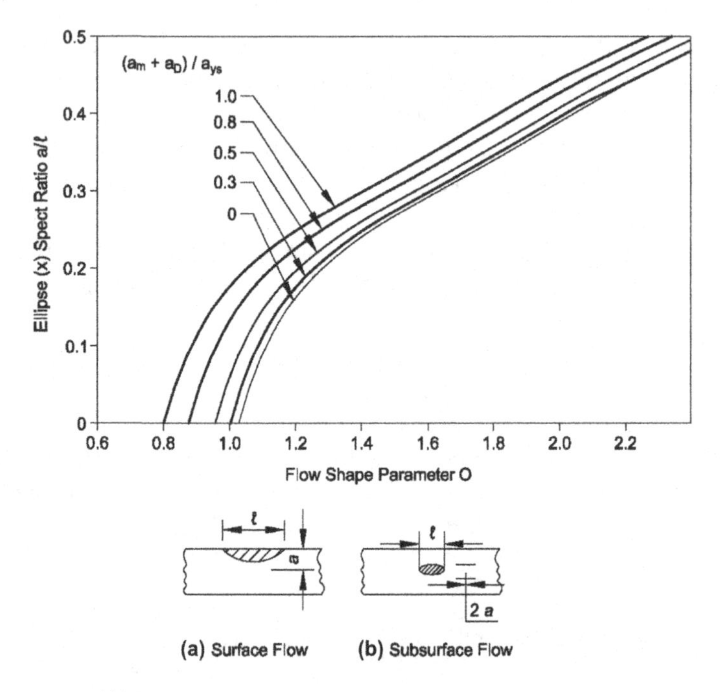

Figure 6.1 Shape factors for flaw model.

D. Determine material toughness and crack propagation properties
   1. Obtain appropriate K_{1C} or J-integral data at ambient and operating temperatures by one of the following:
      a. Testing of samples of the actual material used for construction (this provides the most accurate data but is rarely obtainable)
      b. Published data for the "generic" material of construction (the lower bound trend curve should be used)

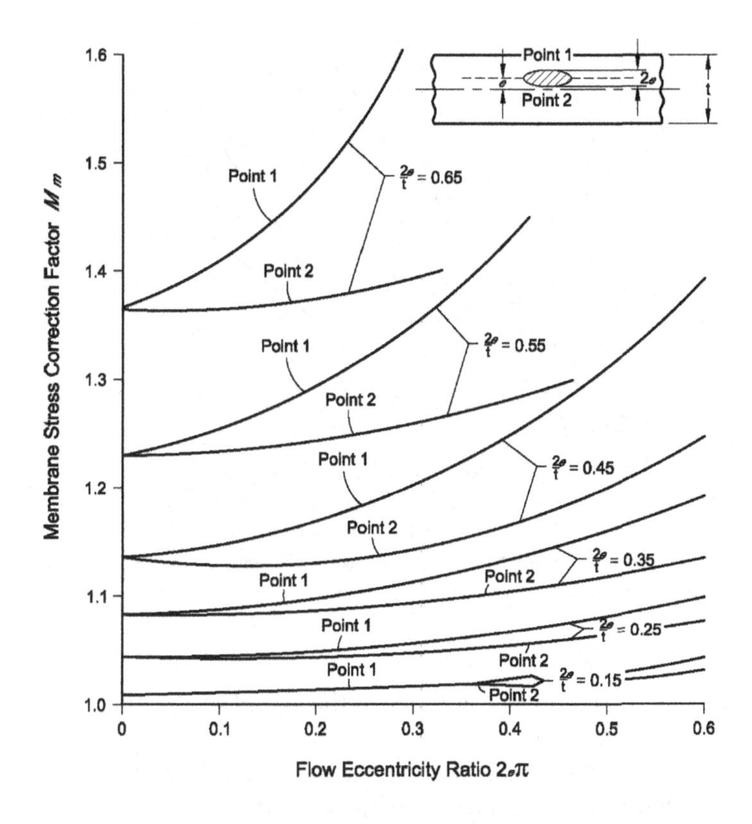

Figure 6.2 Membranae stress correction factor for subsurface flaws.

c. Estimate from C_V impact data for the actual or "generic" material of construction

2. If the flaw is subjected to internal pressure or temperature fluctuations, obtain crack growth rate data by one of the following:

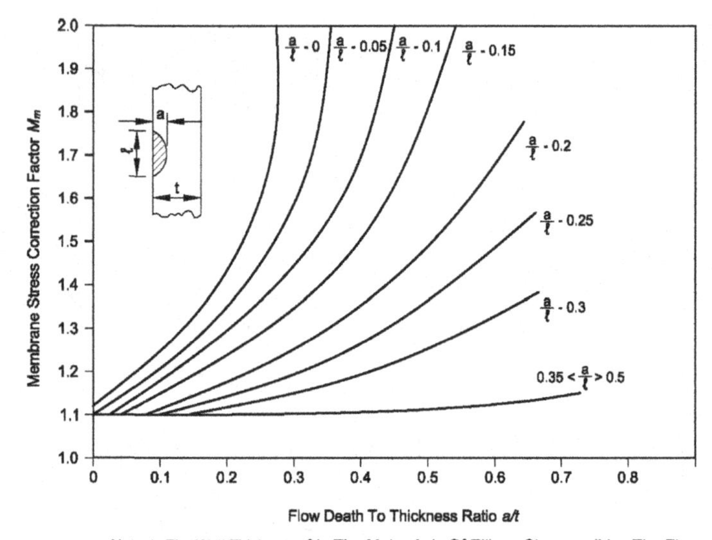

Figure 6.3 Membranae stress correction factor for surface flaws.

   a. Testing of samples of the actual material used for construction
   b. Published data for the "generic" material

E. Determine the acceptability of the flaw
   1. When the flaw is not subjected to significant pressure or temperature cycles (greater than 20% of design pressure or 50°F), compare the stress intensity at the flaw (K_I) with the critical stress intensity for fracture initiation (K_{IC} per item D above) at ambient and operating temperatures:
      a. The flaw is acceptable for operation if K_I is less than K_{IC} at the operating temperature; the flaw is unacceptable for continued operation if K_I exceeds K_{IC} at the operating temperature

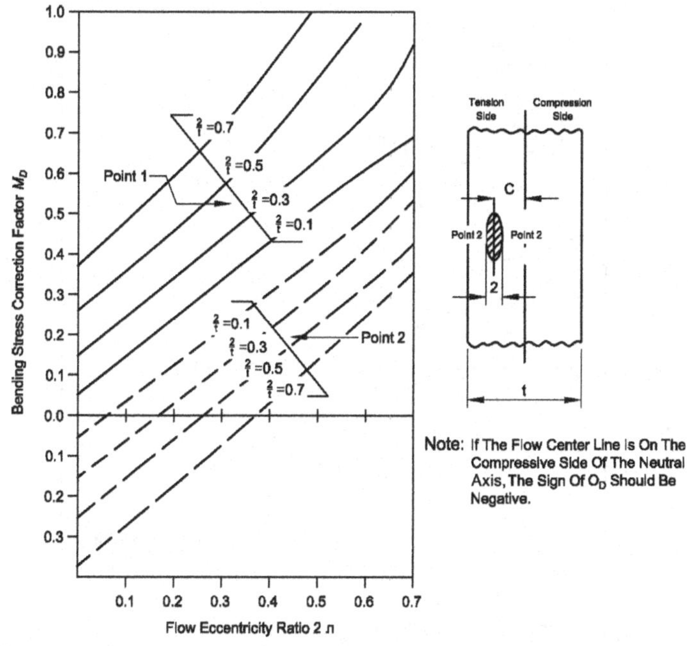

Figure 6.4 Bending stress correction factor for subsurface flaws.

   b. The flaw is acceptable for startup and shutdown at full operating pressure if K_I is less than K_{IC} at ambient temperature; the flaw is unacceptable for startup and shutdown at full operating pressure if K_I exceeds K_{IC} at ambient temperature
2. If the flaw is subjected to significant pressure or temperature cycles, determine the extent of crack growth by fatigue until the next scheduled shutdown
   a. Estimate the number of pressure or temperature cycles until the next scheduled shutdown
   b. Determine the extent of crack growth for the number of anticipated cycles for the stress

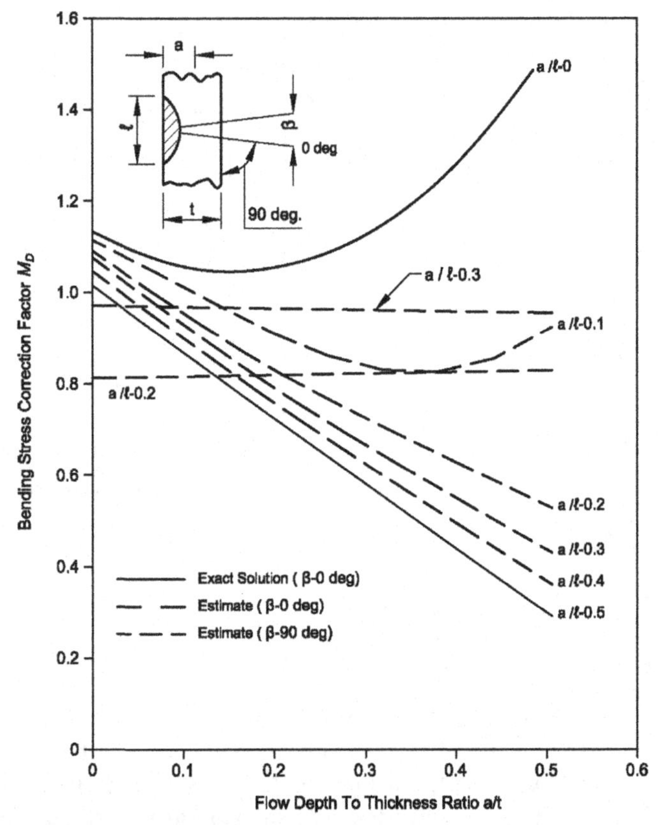

Note: The Wall Thickness: ℓ Is The Major Axis Of Ellipse Circumscribing The Flow

Figure 6.5 Bending stress correction factor for surface flaws.

intensity at the flaw (K_I) using the fatigue crack growth properties (per D)

c. The flaw is acceptable for continued operation if it will not grow to a size such that K_I exceeds K_{IC} prior to the next scheduled shutdown

Required Inspection at the Next Scheduled Shutdown

Flaws that are allowed to remain in a pressure vessel during continued operation based on a fitness-for-service analysis should be reinspected by UT examination during the next scheduled shutdown

If the flaw has grown, the fitness-for-service analysis should be repeated for the enlarged flaw size

If no crack growth is detected, the need for subsequent reinspection should be evaluated by an experienced materials engineer

▶ NONDESTRUCTIVE EXAMINATION (NDE) TECHNIQUES

General Considerations

The following NDE techniques are the most useful for the in-service examination of pressure vessels:

- Visual examination (VT)
- Dye penetrant examination (PT)
- Magnetic particle examination (MT)
- Ultrasonic examination (UT)
- Radiographic examination (RT)

Each technique employs a different phenomenon to interact with flaws in metal objects

Selecting the proper NDE procedures requires understanding each one with regard to the forms of deterioration that may have occurred during operation

Visual Examination (VT)

Should be employed, to some extent, on every vessel

Identifies many forms of deterioration such as the following:

- Corrosion

Erosion
Hydrogen blistering
Occurrence of large surface cracks

Highlights locations where other forms of deterioration may have occurred and will require examination with other NDE techniques

Usually the first NDE performed and will give a useful "first look" of the general condition of the vessel

It alone cannot verify the integrity of a vessel because some forms of deterioration cannot be seen, such as the following:
Fine fatigue or stress-corrosion cracks
Occurrence and growth of internal cracks
Hydrogen attack

It is usually necessary to clean pressure vessels internally to facilitate inspection

Steam or chemical cleaning is used to remove hydrocarbon liquid films and sludge deposits

Hydro-blasting or abrasive blasting is used if corrosion is severe and thick scale adhere to the vessel shell

Nozzles and shell surfaces obscured by internals should not be overlooked

Inspectors frequently use hammers, picks, and scrappers to aid VT in determining the following:
Thickness of corrosion scales
Soundness of the shell under the scale

Depth gauges are used to measure the depth of corrosion pits, and calipers are useful for measuring the inside diameter of nozzles to determine corrosion loss

Whenever it appears that corrosion could have thinned the shell to less than the required minimum thickness, UT examination should be employed

Can be relied on for detecting hydrogen blisters and is useful in appraising the integrity of corrosion protection claddings

However, UT should be used to determine the remaining shell thickness when blisters or deterioration claddings are visually detected

Dye-Penetrant Examination (PT)

General Considerations

Used to detect flaws that are open to the surface of a workpiece

Simple procedure that requires minimal investment

Used primarily to detect cracks that have developed at, or have propagated to, the surface of the vessel shell

ASME Code, Section V, Article 6:

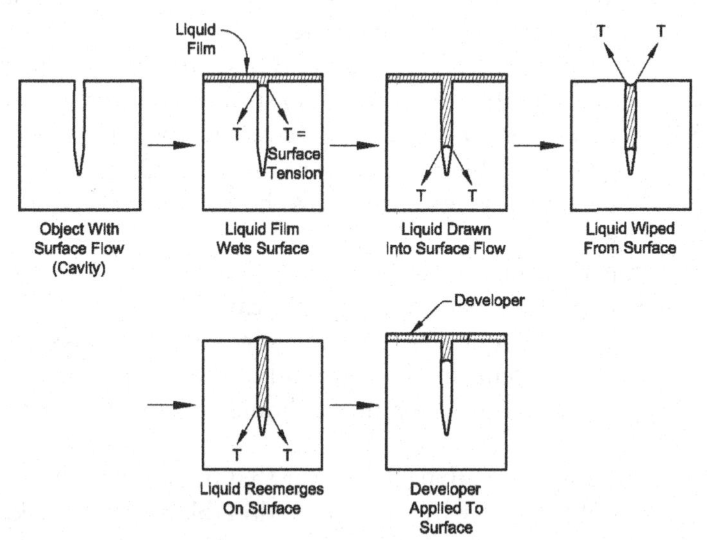

Figure 6.6 Physical principles of dye-penetrant examination (PT).

- Provides minimum requirements for a PT procedure for pressure vessels
- Refers to ASTM Standard SE 165 for additional details that should be considered when establishing a PT procedure

Physical Principles

- Figure 6.6 illustrates the physical principles of PT
- Liquid penetrant is applied to the surface being examined and is drawn into the cavities by capillary action
- The surface tension of the liquid draws it from the wet surface into dry cavities
- A dye (commonly red) is added to the liquid to make it clearly visible
- The surface is then wiped clean of the liquid, which causes a small quantity of the liquid that was drawn into the cavity to reemerge on the surface
- Reemergence onto the surface is due to reverse capillary action, resulting from the now dry surface and a wet cavity
- Sufficient liquid may reemerge to be visible when the surface is wiped clean, but a developer is usually applied to the surface, which greatly enhances the ability to detect the cavity
- Developer (a fine powder):
  - Draws liquid penetrant out of the cavity
  - Forms an opaque layer on the surface that masks the potential confusing background of the workpiece
  - Applied to the surface after the liquid penetrant is wiped clean
  - Powder forms a spongelike layer that draws more liquid out of the cavity by capillary action (blotting), to make it more visible

Color of the developer (usually white) is chosen to contrast with the dye in the liquid to further enhance the visibility of the liquid reemerging from the cavity

Dye-Penetrant Examination Systems

Referred to as the solvent-removable system

Employs an oil-based penetrant that is removed from the surface with a solvent

Both the penetrant and solvent can be sprayed onto the surface with aerosol cans

Surface of the workpiece must be accessible, but there is essentially no other limitation with regard to size, location, and orientation of the workpiece

Surface Preparation

Surface must be clean and dry to enable the liquid penetrant to wet the surface completely and to be drawn into flaws

Condition achieved by washing the surface with a solvent to remove oil, grease, sludge, deposits, and so on

Light grinding to remove surface irregularities can be beneficial; heavy grinding should be avoided, because surface deformation can cover flaws and make it difficult for the penetrant to enter

Corrosion scales can block entry of the penetrant into the flaw and must be removed

Wire brushing, or light abrading, may be adequate for relatively thin scales

Abrasive blasting may be required for heavy scales

Corrosion scales can also fill the cavity and reduce the quantity of penetrant that can be drawn into the opening, which can make surface cracks difficult to detect with PT

Dye Penetrants

Liquid used for a dye penetrant must both wet the surface and have a high surface tension to cause the liquid to be drawn into the cavity

Liquids with a high viscosity flow too slowly to be certain that they will fill cavities in a reasonable time to make them useful penetrants

Can be applied to a surface that has any orientation, by spraying from an aerosol can

Dwell time (allows penetrant to be drawn into flaws):
- ASME, Section V, requires a minimum of 10 minutes
- A dwell time of 20 minutes increases sensitivity for detecting very fine surface cracks

Excess penetrant must be completely removed from the surface before the developer is applied

Wipe off excess with a cloth followed a second time with a clean cloth dampened with a solvent

Surface can be flushed by direct application of the solvent, but the danger of inadvertently removing the penetrant from the flaws before the developer reveals them exists

ASME Code, Section V, prohibits removal by flushing with solvents

Developers

Applied to the surface after the excess liquid penetrant is removed from the surface

Should be applied in a layer thick enough to completely mask the surface, but if the layer of developer is too thick, the liquid penetrant drawn out of a flaw may not reach the top of the layer and the indication of a flaw can be obscured

ASME Code, Section V, sets the minimum development time as 7 minutes

The dye penetrant drawn out of a flaw by the developer should stain the developer layer a deep red

A light pink indicates that excessive cleaning removed penetrant from the flaws and that smaller flaws may not have been revealed

When pale indications are obtained, the examination should be repeated with less vigorous cleaning

Applications and Limitations

PT can be used to detect cracks that have resulted from the following causes:

Stress-corrosion (environmental) cracking

Creep

Mechanical or thermal fatigue

Hydrogen attack

PT will only detect flaws on the surface of the vessel shell

PT does not give any indication of the depth of a surface crack, which is the primary characteristic of a crack that can affect vessel integrity

Usually necessary to determine the size of cracks detected by PT by other NDE procedures, such as UT

Magnetic Particle Examination (MT)

General Considerations

Used to detect surface and near-surface flaws in ferromagnetic materials

Cannot be used for austenitic stainless steels because these materials cannot be magnetized

Similar to PT, in that it is most useful for the detection of surface flaws, but MT is more sensitive in the detection of fine cracks

More difficult to apply than PT, and it requires considerably more investment in equipment

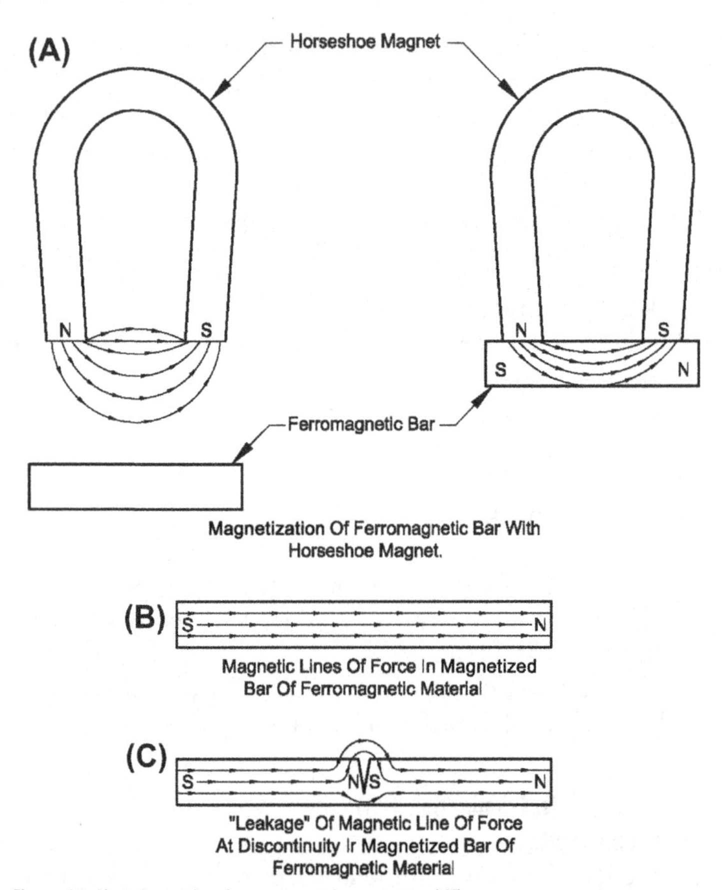

Figure 6.7 Physical principles of magnetic particle examination (MT).

Section V, Article 7, gives the minimum requirement for an MT procedure and refers to ASTM Standard SE 709 for additional details

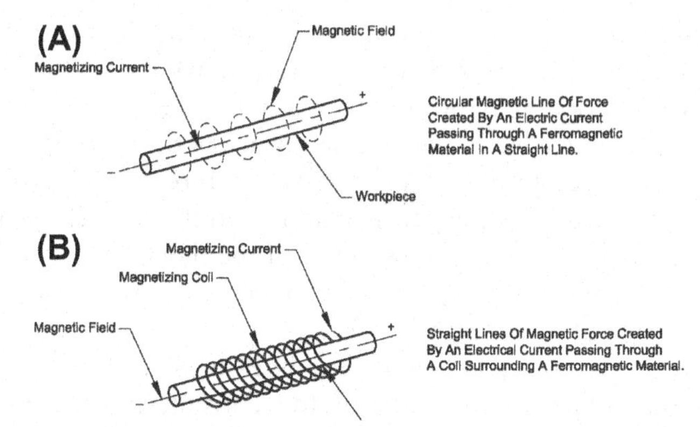

Figure 6.8 Creation of magnetic lines of forces by an electric current.

Physical Principles

A bar of ferromagnetic material can be magnetized by placing it in contact with the north and south poles of a horseshoe magnet, as shown in Figure 6.7A

North and south poles are created in the magnetized bar opposite to the poles of the horseshoe magnet, and magnetic lines of force flow through the bar from the south pole to the north pole in a straight line, as shown in Figure 6.7B

A discontinuity (flaw) in the material will disrupt the magnetic lines of force flowing through the material, as shown in Figure 6.7C

New north and south poles are created at the discontinuity, and some of the magnetic lines of force "leak" from the surface of the material to bridge the gap resulting from the discontinuity

Large leakage occurs in a flaw that is open to the surface

Less leakage occurs in a flaw that is below the surface

Leakage produced decreases as the gap that must be bridged widens

Flaws that are perpendicular to the magnetic lines of force cause the greatest leakage, whereas flaws parallel to the lines of force may not produce any leakage

Magnetic particles applied to the surface of the material will be attracted by the magnetic lines of force leaking from the surface at a flaw and will adhere to the surface at this location, revealing the presence of the flaw

Ferromagnetic materials can also be magnetized by the passage of an electric current

Electric current creates magnetic lines of force that are at right angles to the flow of the current, as shown in Figure 6.8

Because detection of a surface flaw does not depend on a liquid being drawn into the flaw, MT can detect flaws that a liquid penetrant cannot enter because they are too light or are filled with corrosion scale

Magnetic Particle Examination Systems

MT system used in vessels employs an electric current to magnetize the area of the vessel shell being examined

The following systems are common:

- Electromagnetic "yoke"
- Electric "prod" contacts

Electromagnetic Yoke

Functions as a horseshoe magnet, with the magnetic lines of force created by an electric coil in the handle

Magnetic lines of force are developed in the workpiece when the poles of the yoke, functioning as a horseshoe magnet, are brought in contact

Magnetic lines of force flow through the workpiece from one pole of the yoke to the other

(A)

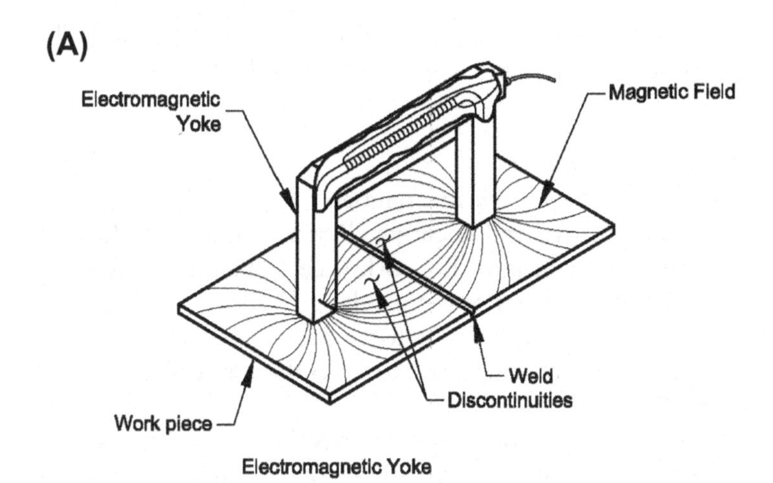

(B)

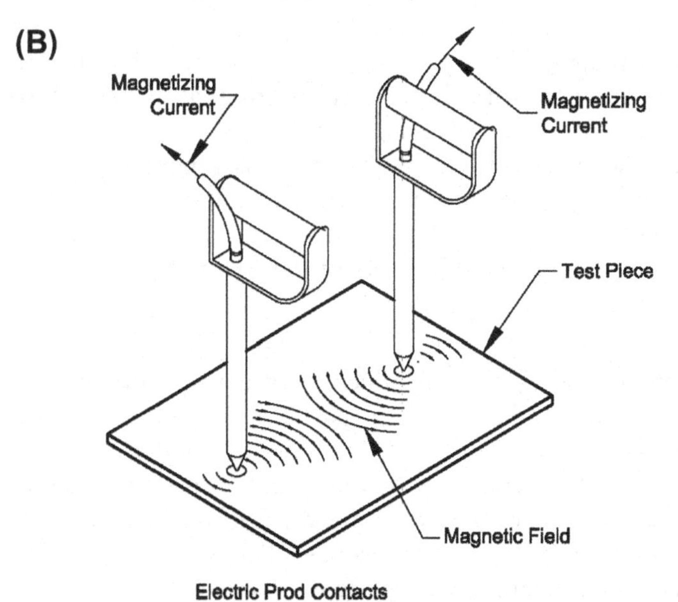

Figure 6.9 Electromagnetic yoke and electric prod contacts for magnetizing the shell of a pressure vessel.

The yoke does not pass an electric current through the workpiece to create magnetic lines of force

Electric Prod Contacts

Do not function as electromagnets

Serve as positive and negative electrodes that introduce an electric current into the workpiece, which flows between the contact points of the prods

Circular magnetic lines of force are created in the workpieces that are concentric around the point of contact of each prod

Prods are often more flexible for MT of components with complex geometries, such as nozzle welds, and are more frequently used than yokes

The direction of the magnetic lines produced by an electromagnetic yoke is different from that produced by electric prod contacts (Figure 6.9)

The greatest sensitivity with a yoke is obtained by positioning its poles normal to the orientation of the flaws, whereas positioning prods parallel to the orientation of the flaws provides the greatest sensitivity

Electric Current

Must be flowing to produce strong magnetic lines of force when the magnetic particles are applied to the surface

When electric prod contacts are used, the current should be turned on after the prods are brought into contact with the workpiece and turned off before the prods are removed from the workpiece to avoid arcing

Arcing can cause small heat-affect zones to develop in the workpiece, similar to welding, which could make the vessel susceptible to failure by stress-corrosion cracking in some process environments

Arcing is not a concern when an electromagnetic yoke is used, because the yoke functions as an electromagnet and does not make electrical contact with the workpiece

AC or DC currents can be used:

- DC currents penetrate farther below the surface and should be used when subsurface flaws must be detected
- AC currents provide greater sensitivity for detecting flaws that are open to the surface, and they are preferred when only surface flaws must be detected

The electric current must create magnetic lines of force in the workpiece of sufficient strength for the magnetic particles, applied to the surface, to be attracted to the magnetic leakage occurring at flaws

ASTM SE 706 recommends a magnetizing current

From 90 to 100 amperes/inch of prod spacing for thicknesses less than {3/4} inch

From 110 to 125 amperes/inch of prod spacing for thicknesses {3/4} inch and larger

Undesirable for the prod spacing to exceed 8 inches

Because a yoke functions as an electromagnet, the criteria for the electric current are different

Magnetizing strength of a yoke is determined by its lifting power

Yoke must be able to lift a 10-pound steel plate when an AC current is used or a 40-pound steel plate when a DC current is used

Magnetic Particles

Two types: wet and dry

Classified according to how they are carried to the workpiece:

Dry particles are carried by air
Wet particles are carried by a liquid
Both dry and wet particles must have a high magnetic permeability so that they can be attracted by relatively low levels of magnetic leakage at flaws

Dry Particles

Applied with low velocity as a uniform cloud (colored yellow, red, or fluorescent)
Used in rough surfaces that inhibit the flow of liquids
Sensitive to detect subsurface flaws
Can be applied to surfaces that are too warm for liquids

Wet Particles

Suspended in light oil or water and applied to the surface by spraying
More sensitive at detecting fine surface cracks
Fluorescent coatings make them highly visible in ultraviolet (UV, black) light, and they adhere to the surface after the liquid has evaporated

Applications and Limitations of MT

MT is preferred to PT for detecting surface cracks because of its greater sensitivity, especially when wet particles are used
MT can detect very fine cracks that escape detection by PT
Especially useful for vessels that have been exposed to service environments that contain H_2S under conditions that can cause stress-corrosion cracking
Similar to PT, MT does not give a reliable indication of the depth of a surface crack; the depth of these cracks can be determined using UT

MT can detect subsurface cracks that are not too far below the surface but do not depend on MR alone to detect internal cracks

UT or RT should be employed to supplement UT

Especially true for thick (greater than 1.5-inch thick) vessel shell components

Ultrasonic Examination (UT)

General Considerations

UT is a versatile NDE method

Able to determine flaw size

Accuracy of the sizing data obtained is dependent on the following:

Procedure used

Qualifications and experience of the technician

ASME Code, Section V, Article 5, gives the minimum requirements for a UT procedure for the in-service examination of a pressure vessel

Minimum requirements may not be adequate to detect and size some types of flaws

Special UT techniques can improve detection and sizing capabilities

Physical Principles

A high-frequency sonic (ultrasonic) wave is introduced into a solid material and is reflected by interfaces in the material

Reflected waves are analyzed to detect the existence of a flaw and determine its location

Interfaces that reflect the ultrasonic waves can be between the solid material and a gas, liquid, or other solid

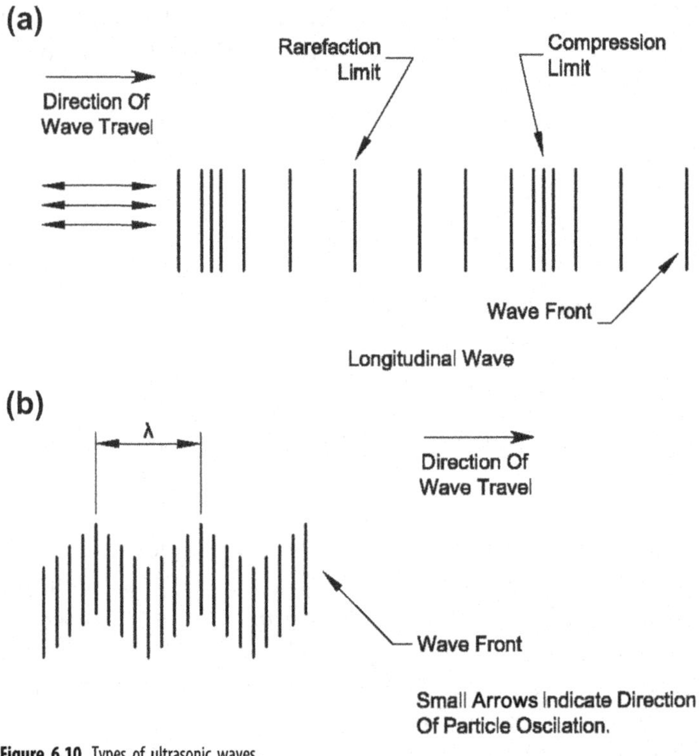

Figure 6.10 Types of ultrasonic waves.

Solid-gas interfaces tend to reflect the waves, whereas solid-liquid and solid-solid interfaces only partially reflect the waves

Flaws in metal vessel shells normally act as either solid-gas or solid-solid interfaces and thus can be either very good or very poor reflectors

Surfaces of the vessel shell are solid-gas interfaces that are normally very good reflectors

There are two types of ultrasonic waves (Figure 6.10):

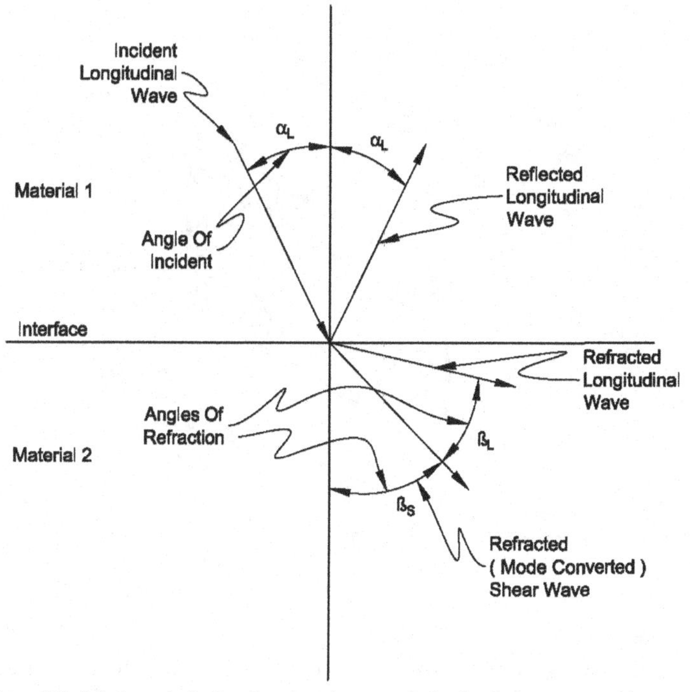

Figure 6.11 Reflection and refraction of an ultrasonic wave at the interface between two materials.

Longitudinal wave: develops when the particles are displaced parallel to the direction of propagation

Transverse (shear) wave: results when particles are displaced perpendicular to the direction of propagation

Transverse (shear) waves will generally propagate only in solids, because the interatomic forces in liquids and gases are too weak for perpendicular displacement to induce the displacement of adjacent particles (Figure 6.11)

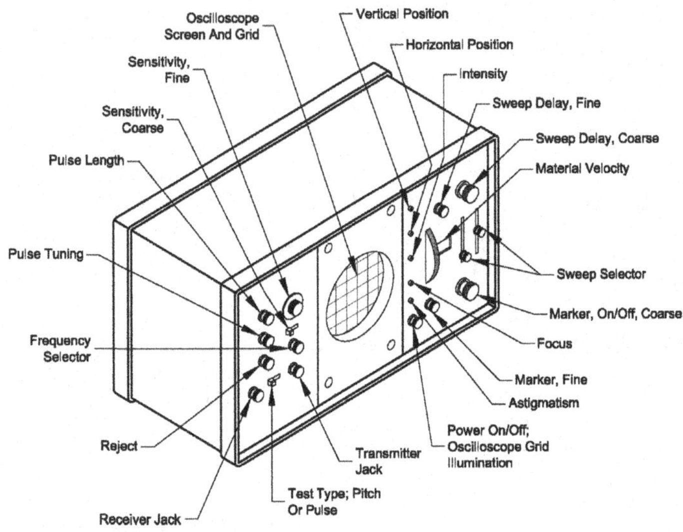

Figure 6.12 Ultrasonic instrument for pulse-echo UT.

Ultrasonic Instruments

Most UT systems employ a "pulse-echo" method of operation, in which the same transducer that creates the ultrasonic wave (pulse) is also used to detect any ultrasonic waves (echoes) reflected by flaws

A typical ultrasonic instrument for the pulse-echo UT is illustrated in Figure 6.12

"Pitch-catch" or "dual-element" transducers have improved the reliability of detecting and accurately sizing certain types of planar flaws (or cracks) that occur during service, such as tight and branched stress-corrosion cracks

Generally, high-frequency electrical impulses that activate an ultrasonic transducer

Frequencies range from 0.5 to 10 MHz

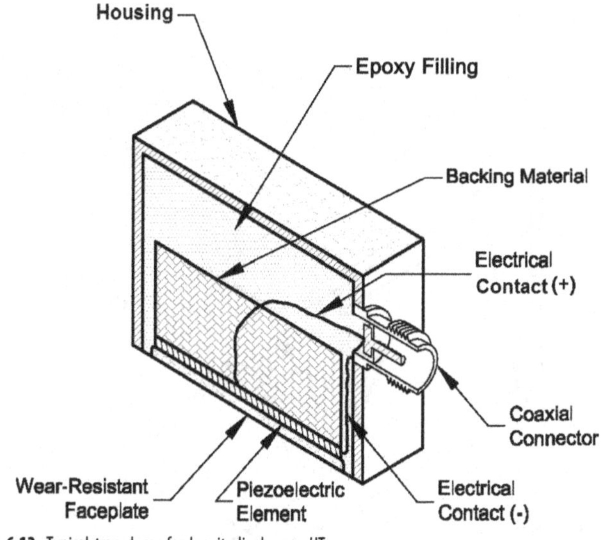

Figure 6.13 Typical transducer for longitudinal wave UT.

Provide greater sensitivity for detecting small flaws, but penetration is reduced because of the greater attenuation attributable to scattering by small irregularities in the material

Metals with coarse-grained microstructures, such as stainless steel, can present a difficult problem for obtaining adequate penetration while retaining sufficient sensitivity for detecting small flaws

Low frequencies can be used to increase penetration, but sensitivity for detecting small flaws will be sacrificed

Controls are provided for setting the duration of the pulses transmitted and the interval between successive pulses

Settings depend on the thickness of the workpiece

Desirable for all echoes of one pulse to fade out in the workpiece before the next pulse is emitted

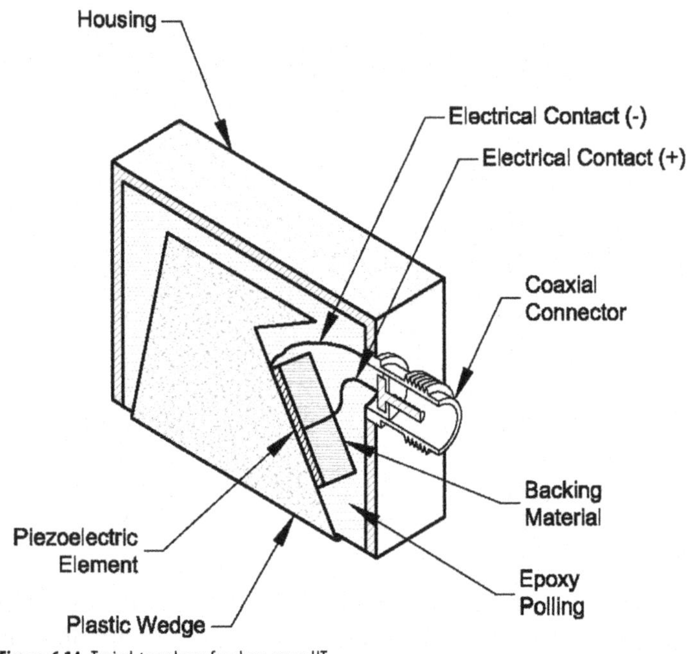

Figure 6.14 Typical transducer for shear wave UT.

An amplifier is incorporated into the instrument to amplify the echoes received by the transducer, and a gain (or sensitivity) adjustment is provided to optimize the signal-to-noise ratio

Ultrasonic echoes received by the transducer are visually displayed on an oscilloscope

The following are displayed:

Amplitudes of the echoes

Times at which they are received

Ultrasonic Transducers

Ultrasonic waves are created in a metal workpiece with a piezoelectric transducer that is coupled to the workpiece

Piezoelectric transducer converts high-frequency electrical impulses into correspondingly high-frequency mechanical vibrations that induce ultrasonic waves in the workpiece

"Longitudinal wave UT":

Ultrasonic wave that is introduced perpendicular to the surface of the workpiece (Figure 6.13)

"Shear wave UT":

Ultrasonic wave that is introduced at an angle to the surface of the workpiece (Figure 6.14)

Couplants

Attenuation of ultrasonic waves propagating through air is very high, and normal surface roughness will always create an air gap between the transducer and the workpiece

Therefore, a couplant is used that will allow the waves to propagate from the transducer to the workpiece with significantly less attenuation

Satisfactory couplants are usually viscous liquids or greases that fill the surface irregularities and thus eliminate the air gap

Rough workpieces may have to be ground to obtain a sufficiently thin layer so as to obtain good transmission

Some surface preparation by light grinding is frequently required when corrosion has roughened the surface or if scales have formed

Longitudinal Wave UT

Used primarily to determine the remaining thickness of corroded shell components, including the depth of pits

Used to detect internal flaws that have developed during service and that have a reflecting surface essentially parallel to the surface of the workpiece, such as hydrogen blisters

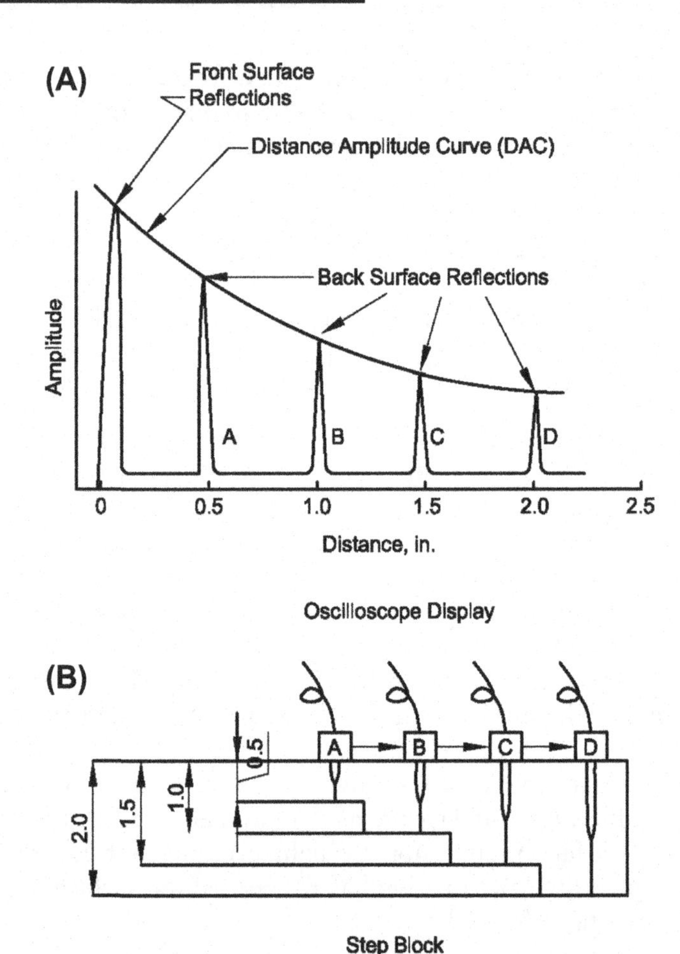

Figure 6.15 Calibration of longitudinal wave UT for thickness measurement.

Calibration

Consists of developing a "distance amplitude curve" (DAC) for the instrument and transducer, using test blocks manufactured from a material similar to the workpiece

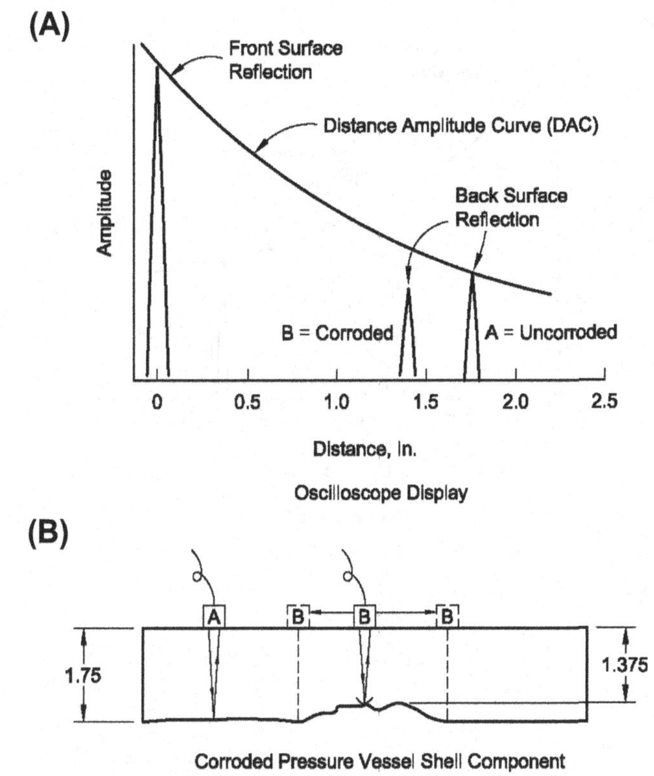

Figure 6.16 Determination of remaining thickness of corroded pressure vessel shell with longitudinal wave UT calibrated in Figure 6.15.

Calibration for determining the remaining thickness of a corroded component can be accomplished by placing the transducer sequentially on test blocks with different known thicknesses (Figure 6.15)

Calibration

The amplitude of the reflection (echo) received by the transducer from the back surface decreases as the

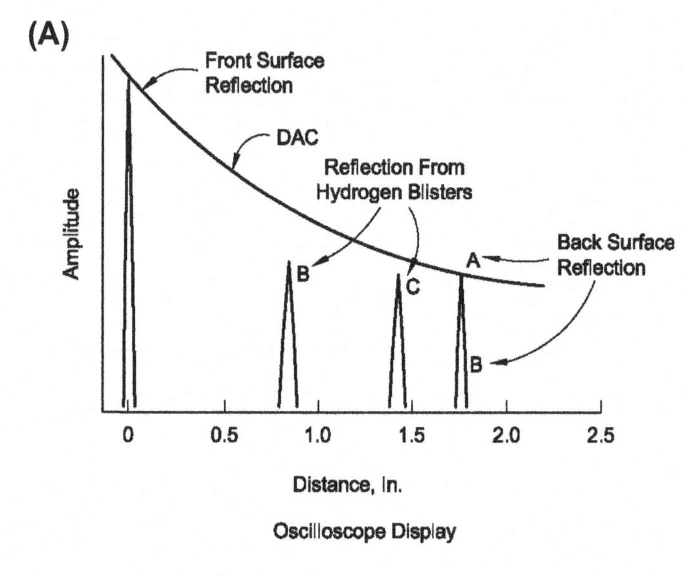

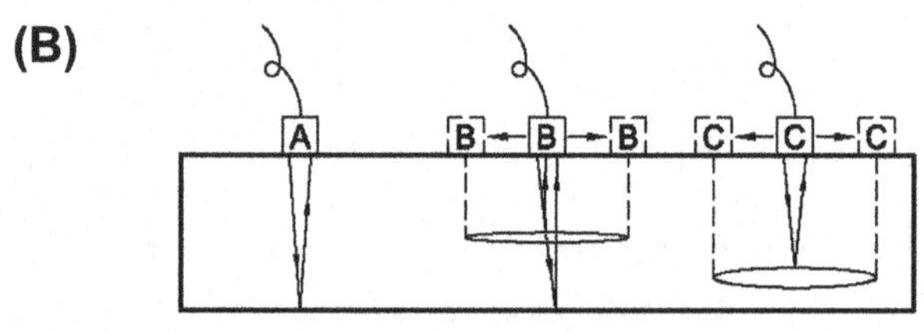

Figure 6.17 Detecting hydrogen blisters with longitudinal wave UT.

thickness of the test block increases because of the attenuation of the ultrasonic wave propagating through the block

Remaining Thickness

Figure 6.16 illustrates the use of longitudinal UT for determining the remaining thickness of a corroded shell

If the transducer is placed on a location of the shell component that is not corroded, as depicted by the position of transducer A in Figure 6.16, a back reflection will be observed at a distance equal to the original thickness of the shell

If the transducer is placed on a location where internal corrosion has occurred, as shown by the position of transducer B in Figure 6.16, a back reflection will be observed that is less than the original thickness

Hydrogen Blisters

Figure 6.17 illustrates the use of longitudinal wave UT for detecting, locating, and determining the size of hydrogen blisters

The blisters are internal flaws that have a reflecting surface at a depth from the front surface that is less than the distance to the back surface (thickness) of the shell component

The blisters will cause reflected peaks to appear in the oscilloscope display at a distance less than the thickness of the shell

The distance of the peaks on the oscilloscope display resulting from the blisters indicates their depth below the surface, and the size of each blister can be estimated by moving the transducer along the surface until the reflection attributable to the blister disappears

Shear Wave UT

Used primarily to detect and determine the size of cracks that have developed during service

Can provide very good data for the evaluation of vessel integrity, but the quality of the data depends on the skill and expertise of the technician performing the examination

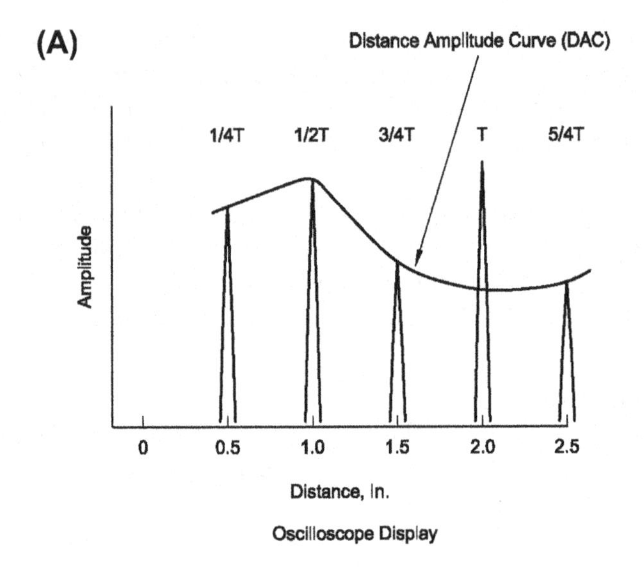

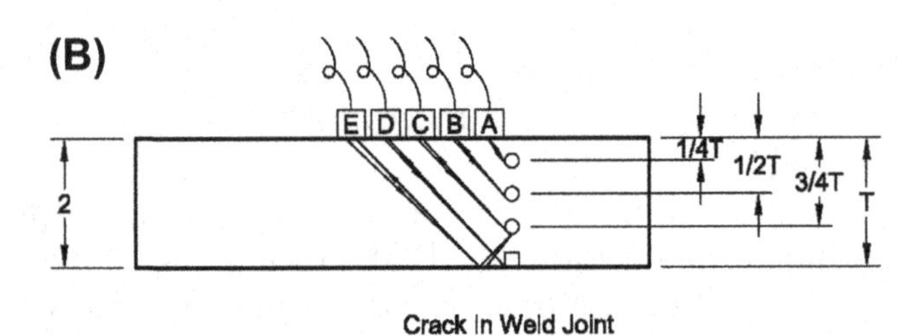

Figure 6.18 Calibration of shear wave UT.

Calibration

Calibration using a test block manufactured from a material similar to that of the workpiece that has side-drilled holes and a notch on the back surface (Figure 6.18)

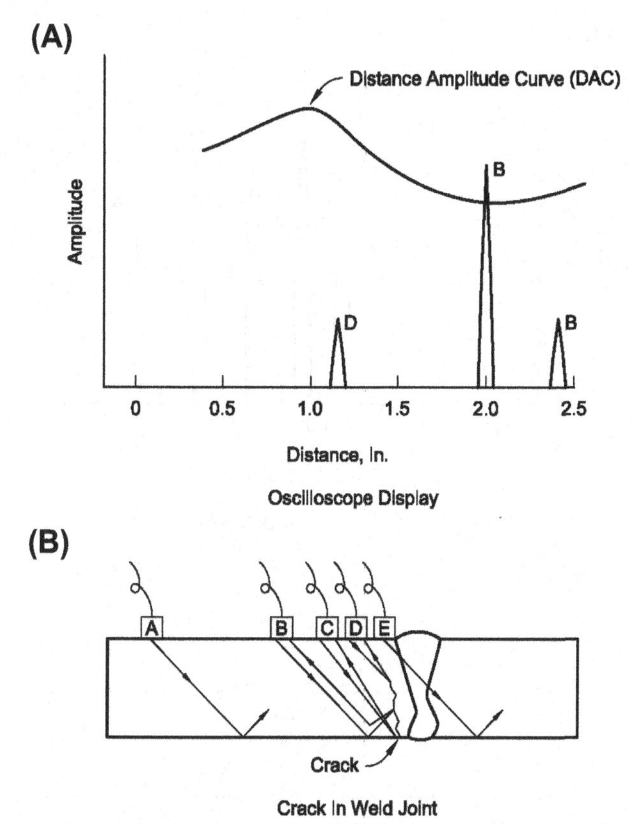

Figure 6.19 Shear wave UT calibrated according to Figure 6.18 for detecting a crack in the heat-affected zone of the weld joint.

The test block should have a thickness within 1 inch of the thickness of the workpiece

The side-drilled hole should be {3/16} of an inch in diameter for test blocks up to 4 inches thick and increase {1/16} of an inch in diameter for each 2-inch increase in thickness of the test block greater than 4 inches

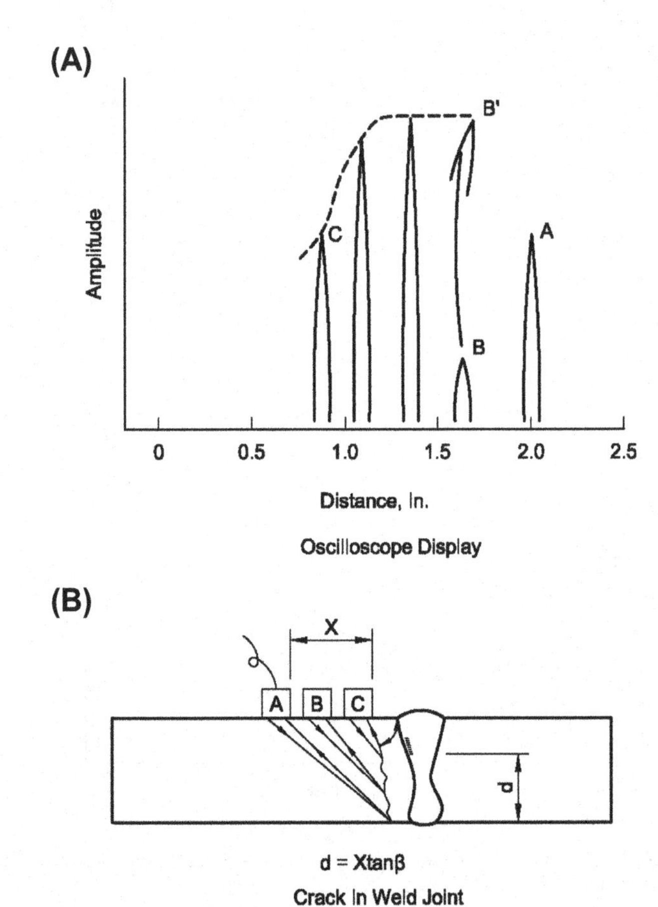

Figure 6.20 Amplitude-based sizing using shear wave UT.

The transducer is moved along the surface until the reflection from the {1/4}-, {1/2}-, and {3/4}-thickness drill holes attain maximum peak amplitude

Crack Detection

Shear wave UT is very useful for detecting cracks that have developed during service

Figure 6.19 illustrates how shear wave UT can be used to detect a crack in the heat-affected zone of a weld joint

This crack has started at the back surface of the workpiece (ID of the vessel) and has propagated toward the front (OD) surface

Shear wave UT is performed by moving the transducer along the OD surface of the vessel toward the weld joint; no reflection is observed in the oscilloscope display with the position of transducer A in Figure 6.19, because the shear wave is reflected by the ID surface away from the transducer

A relatively strong reflection from the base of the crack will be observed in the oscilloscope display for the position of transducer C, because the base of the crack at the ID surface acts like a notch in the back surface of the test block

Crack Sizing

UT crack sizing techniques are classified as follows:

- Amplitude-based
- Time-based

Amplitude-Based Sizing

Most frequently used

Figure 6.20 illustrates how the depth of a crack is determined using amplitude-based sizing

The crack was detected by shear wave UT as shown in Figure 6.19

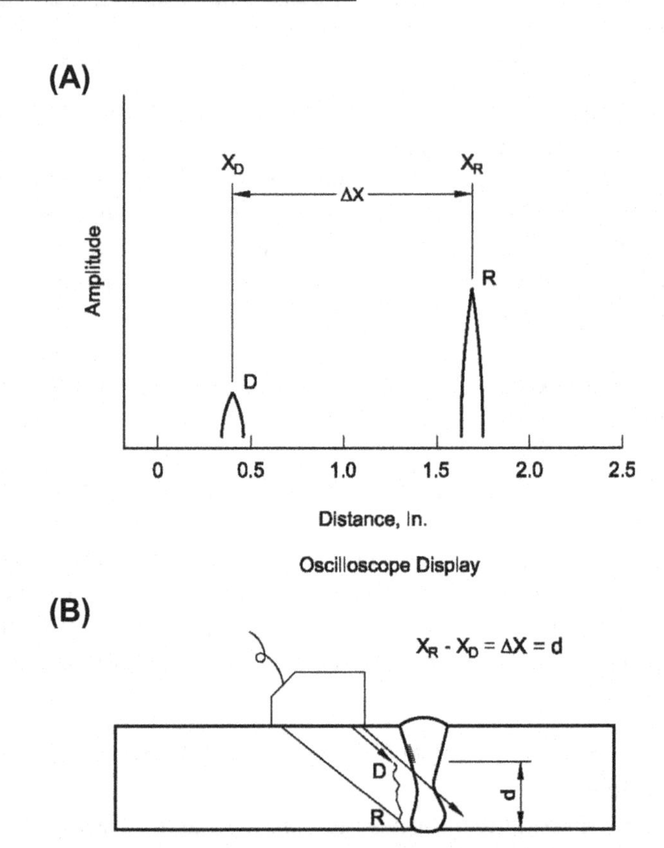

Figure 6.21 Time-based sizing using shear wave UT.

The occurrence of a relatively high amplitude peak in the oscilloscope display at a distance corresponding to the thickness of the shell confirms that the crack initiated at the ID surface

Offers satisfactory results for crack depths that exceed the diameter of the transducer

Not satisfactory for cracks shallower than the diameter of the transducer; this is the situation with most vessel inspections

A {1/8}-inch-deep crack may not affect the integrity of a vessel, whereas a {3/4}-inch-deep crack would require repair before the vessel is returned to service

Amplitude-based sizing tends to oversize small cracks and undersize large ones; therefore, it may not accurately discriminate between the shallow superficial crack and the deeper one that could jeopardize safe operations

Time-Based Sizing

Generally provides greater accuracy for determining depth, especially when the crack is relatively small with respect to the diameter of the transducer

Thus, it should be used whenever a fitness-for-service analysis is made to evaluate the integrity of a vessel

Figure 6.21 illustrates one shear wave UT technique that can be used to determine the depth of a crack

The wave propagates through the material as a wave front that has a width at the surface equal to the diameter of the transducer

Two peaks, referred to as a doublet, are observed in the oscilloscope display for the position of the transducer

A peak with relatively high amplitude (peak R) will be developed by the corner reflection of the wave from the base of the crack at the ID surface; this peak will have maximum amplitude at the distance in the oscilloscope display corresponding to the thickness of the vessel shell

The position of the wave that passes over the top of the crack is diffracted, which forms the second peak (peak D) with a lower amplitude

The separation between the reflected and diffracted peaks is constant whenever they are observed together, regardless of the position of the transducer

The distances in the oscilloscope display of the separation of the peaks indicate the depth of the crack (Δ in Figure 6.21)

Applications and Limitations

UT is a very efficient NDE method

Longitudinal Wave UT

Determines the remaining wall thickness of a corroded pressure vessel under almost any circumstances

Detects and locates hydrogen blisters or similar internal flaws

Shear Wave UT

Useful for detecting cracks and provides essentially the only method for determining the size (depth) of cracks with sufficient accuracy for making a fitness-for-service analysis to evaluate the integrity of a vessel

Limitations

Accuracy of the data obtained greatly depends on the skill and expertise of the technicians performing the examinations; this is especially true for detecting and sizing cracks by shear wave

Because UT is not used during construction, many indications of flaws detected by UT are difficult to classify as either innocuous fabrication flaws or more serious indications of NDE techniques

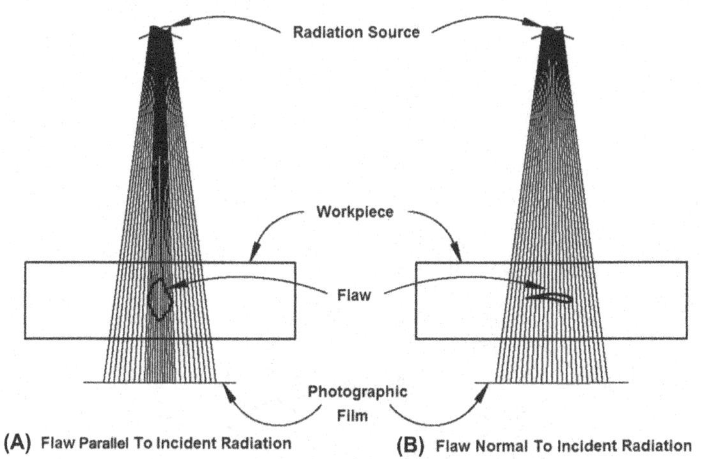

Figure 6.22 Attenuation of radiation by a workpiece containing a flaw.

Radiographic Examination (RT)

General Considerations

RT is useful for detecting both surface and internal flaws

Primary NDE procedure required by the ASME Code to verify the quality of welds during construction

Associated radiation hazard makes it difficult to use for inspections of pressure vessels during shutdowns, when other personnel are working on or near the vessel

Investment in equipment (radiation sources and darkroom facilities for processing film) can also be quite high, but this can be offset by the use of qualified contractors

ASME Code, Section V, Article 2, gives the minimum requirements for an RT procedure for pressure vessels

Physical Principles

X-rays are gamma rays that penetrate steel, but the intensity of the incident radiation will be attenuated as it passes through the material

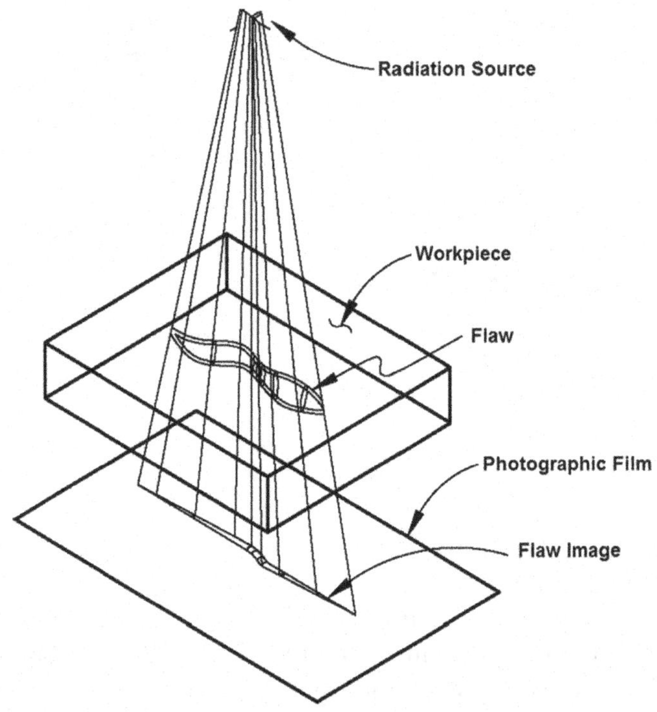

Figure 6.23 Projection of three-dimensional flaw in a workpiece onto two-dimensional surface of film.

Degree of attenuation depends on the thickness and density of the material

Flaws can have the effect of reducing the thickness of material through which the radiation must pass by interposing cavities or impurities of lower density in the workpiece

Thus, there is less attenuation of radiation passing through the flaw than through the surrounding material (Figure 6.22)

Photographic film placed opposite the source of radiation will be exposed by the radiation that has passed through the workpiece; consequently, the flaw will appear as a dark image on the developed negative (referred to as a radiograph)

The radiation passing through the workpiece does not directly interact with the flaw

The flaw is detectable only because it alters the thickness of material through which the radiation passes

The image of the flaw on the radiograph is actually the silhouette of the three-dimensional flaw projected onto the two-dimensional surface of the film (Figure 6.23)

Flaws will not always reduce the thickness of material through which the radiation must pass

The flaw in Figure 6.22B is identical to the flaw in Figure 6.22A, except that it is rotated 90 degrees

With this orientation, the flaw will not reduce the thickness of the material

Thus, the attenuation of radiation passing through this flaw will be essentially the same as that for the radiation passing through the surrounding material, and there will be no indication of the flaw

Radiation Sources

Both X-rays and gamma rays are used as the incident radiation

Energy and intensity are the most important characteristics of incident radiation

- Energy of incident radiation determines its ability to penetrate the workpiece

Thicker workpieces require higher-energy radiation, but it reduces the scatter of radiation passing through the workpiece

Intensity of the radiation reaching the photographic film after it has passed through the workpiece controls the length of time required to properly expose the film

Higher-intensity radiation is required for thicker workpieces to obtain reasonable exposure times, as attenuation of the radiation passing through workpiece increases with material thickness

Radiation sources used for RT have much higher energies and intensities than those used for medical X-rays; thus, special safety precautions should be observed (shielding, etc.)

Guidelines for shielding and restriction of access should be obtained from knowledgeable safety and health specialists before performing the RT

Photographic Film

The radiation that passes through the workpiece is recorded by photographic film

The radiographs are usually interpreted visually with the aid of a high-intensity light source (light box), but optical densitometers or image analyzers are occasionally used

Three primary characteristics of the film can affect the sensitivity of RT for detecting flaws:
- Gradient
- Grain
- Speed

Gradient

The difference in optical density of the negative resulting from exposure by different intensities of radiation

A high gradient results in a relatively high-contrast negative, which makes small differences in the intensity of radiation passing through the workpiece visible

Films that have a high gradient provide the greatest sensitivity for detecting small flaws that cause only a small attenuation of the radiation

Grain

Results directly from photosensitized crystals in the film

Darkened crystals impart a visually apparent "graininess" to the transparent negative that limits the detail that can be resolved by viewing the negative

The grain of the film can obscure very fine flaws; thus, the flaws may not be visible in a radiograph

Fine-grain films usually also have a high gradient and are preferred for RT to obtain the greatest sensitivity for the detection of flaws

Speed

A high-speed film requires less exposure to radiation to produce the same optical density in the developed negative than a low-speed film

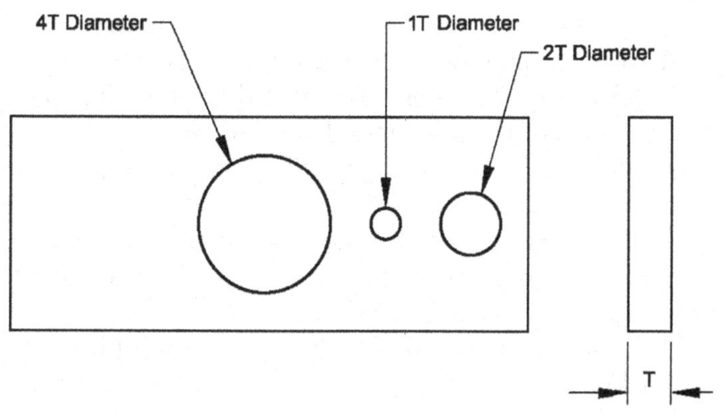

Figure 6.24 Hole-type IQI for evaluating the quality of a radiograph.

Therefore, less exposure time is required for high-speed films to produce satisfactory radiographic films than with low-speed films for the same intensity of incident radiation

Exposure

The exposure of the photographic film by radiation passing through the workpiece is determined by the intensity of the radiation multiplied by the time of the exposure

ASME Code, Section V, requires radiographic film to have a density between 1.8 and 4 for proper visual interpretation

The exposure must be adjusted if the density of the radiograph is not between these limits, by changing either the time of the exposure or the intensity of the incident radiation

Radiograph Quality

The acceptability of a radiograph for the detection of flaws is determined with a device, referred to as an "image quality indicator" (IQI), that is placed on the surface of the workpiece when the exposure is made

Hole-type IQIs are relatively thin pieces of material with radiation attenuation characteristics similar to the workpiece that contains holes with diameters that are 1, 2, and 4 times the thickness (Figure 6.24)

The quality level required for a radiograph is designated by a two-part expression, X–YT:

- "X" is the maximum thickness permitted for the penetrameter as a percentage of the thickness of the workpiece
- "Y" is the diameter of the hole as a multiple of the thickness
- "T" is the thickness

A quality level of 2 to 2T is adequate for most applications of RT for the in-service inspection of pressure vessels and is consistent with the requirements of ASME Code, Section V

A radiograph is considered to be acceptable for the quality specified if the entire outline of the penetrameter is visible, the density of the penetrameter is within the required range of 1.8 to 4, and the hole is discernable

Applications and Limitations

The reliance of the ASME Code upon RT should not be construed to imply that it is the optimum NDE method to use for the in-service inspection of vessels

In fact, most flaws that can develop as a consequence of the deterioration of a vessel during service can be better detected by other NDE methods

One circumstance where RT can be within considerable advantage is when a direct comparison is desired between the present condition of a vessel and its condition when new, and other NDE methods were not used during construction to provide baseline data

Limitations of RT for in-service inspections

- It will only detect cracks that are essentially parallel to the direction of the incident radiation and have a sufficient width to be visible in the radiograph
- Does not give a reliable indication of the depth of a flaw through the shell

RT is also severely limited for the in-service inspection of nozzle openings and welds, which tend to be in locations of relatively high stress

RT is the most time consuming and expensive of all NDE procedures

Additional time and cost penalties are incurred indirectly by restricting and delaying other work in the area because of the serious radiation hazard associated with RT

Therefore, the suitability of other NDE methods for detecting the forms of deterioration that might have occurred during service should be investigated before RT is employed

RT can provide valuable data concerning the integrity of a vessel; it also provides permanent records that can be compared to the results of future inspections or reinterpretation in light of new information concerning the deterioration that can occur during service

Repair, Alteration, and Re-rating

7

▶ OVERVIEW

This chapter discusses the following topics:

Repair, alteration, and re-rating of pressure vessels
Pertinent code and jurisdiction requirements
Differences among
ASME Code
National Board Inspection Code
API 510
Design of repairs
Planning and approval
Responsible organization
Materials
Replacement parts
Post-weld heat treatment
Inspections and hydrotest procedures
Documentation and nameplates

Inspection of a pressure vessel frequently reveals that some form of deterioration has occurred during service

Analysis of the deterioration may indicate that

- The vessel was repaired under the original design conditions
- The vessel was re-rated for less-severe design conditions

Re-rating for new design conditions may also be necessary because of changes in operating requirements

New process requirements may also lead to alterations to a vessel

The size and location of a vessel in a facility/plant can make repair or alteration difficult
Non Destructive Examination (NDE) of repairs and alterations is very important to assure that high integrity has been obtained
More extensive NDE than was required for the original construction is usually advisable

▸ CODE AND JURISDICTION REQUIREMENTS

ASME Code

Applies directly only to the "design, fabrication and inspection during (the original) construction of pressure vessels"
Applicability terminates when the authorized inspector authorizes application of the code stamp
The ASME Code should not be interpreted to imply that all design details and fabrication procedures that are not covered by its rules are unsatisfactory for repairs and alterations
The ASME Code is formulated around design details and fabrication procedures that are obtainable with good shop practices
Jurisdictional requirements
- Most authorities having jurisdiction require the owners/operators to operate and maintain their vessels in a safe condition
- The majority of these authorities have established regulations that refer to either
  - The National Board (NB) Inspection Code
  - The American Petroleum Institute (API) Pressure Vessel Inspection Code (API 510) for the repair, alteration, and re-rating of pressure vessels

Jurisdictional Requirements

Most companies prefer to use API 510 because it is specifically oriented to the needs of the hydrocarbon processing industry

Some authorities having jurisdiction make the owner/operator responsible for obtaining approvals and filing the documentation for repairs, alterations, and re-rating

National Board Inspection Code versus API 510

Technical requirements are similar

Procedural and administrative aspects differ

Major differences are that the National Board Code does the following:

- Requires an authorized inspector to hold a commission from the National Board
- Restricts the authority of an authorized inspector employed by the owner/operator
- Requires preparation and approval of an R-1 Form and the attachment of a new nameplate for repairs and alterations that do not change the maximum allowable working pressure or design temperature

The more elaborate procedural and administrative details of the National Board do not result in repairs and alterations with higher integrity, but they can considerably increase the costs incurred

Figure 7.1 is a flowchart that compares the major requirements of API 510 to those of the National Board

API 510

Permits greater flexibility through the exercise of engineering judgment by the owner/operator than is possible when following the National Board

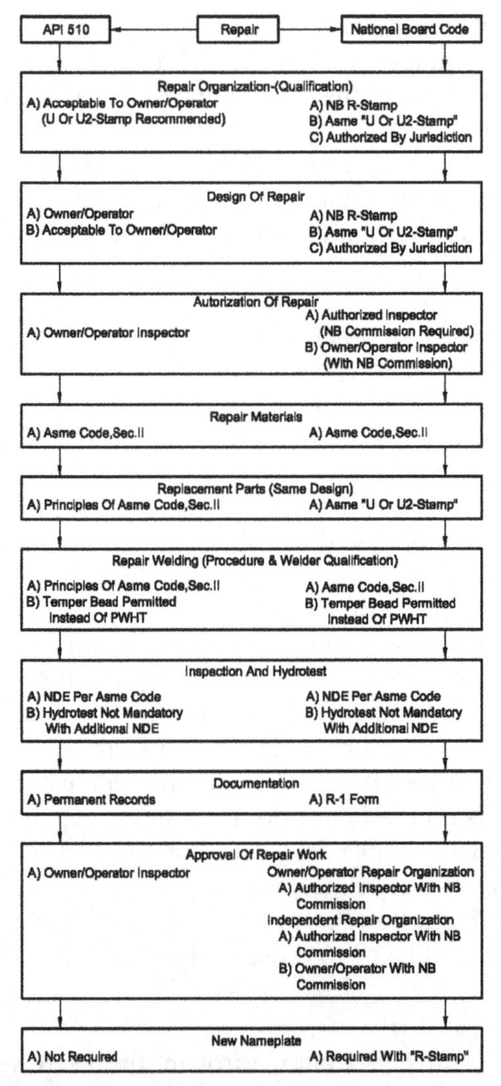

Figure 7.1 "Flow Chart for Repair Work Requirements"

Owner/operator has more responsibility for the integrity of a repair

References to the ASME Code

Both refer to the ASME Code for making repairs, alterations, and re-ratings of pressure vessels

Wording used by the National Board and API has different implications

Requirements of API 510 versus the National Board Inspection Code

National Board

Requires all repairs and alterations to conform to the ASME Code whenever possible (Paragraph R-100)

API 510

Requires following the principles of the ASME Code

Both codes recognize that it may not always be possible to adhere strictly to the ASME Code

The National Board implies that the code must be complied with whenever possible

The API permits more flexibility for deviation from the code by exercising engineering judgment

Design details for repairs and alterations that deviate from the rules of the code should be justified by a stress analysis to verify that the maximum allowable stress permitted by the code is not exceeded

Fabrication procedures that differ from the original construction must be properly qualified to verify the following:

- The minimum materials properties (strength and C_v impact toughness) required by the code are obtained,
- Any other materials requirements specified for the service conditions (such as maximum hardness of the weld metal and heat-affected zones) are achieved

Authorizations and Approvals

Both the NB and the API require obtaining authorizations from an "authorized inspector"

NB requires that inspectors hold a commission from the NB (Chapter 1)

API requires only that the inspector be qualified to perform the inspection (Paragraph 4.2.4) by virtue of the inspector's knowledge and experience (many company inspectors are commissioned by the NB)

NB's policy:

Emphasizes compliance with its rules through scrutiny of an authorized inspector, consistent with its dictum to conform to the ASME Code

API's policy:

- Emphasizes compliance
- Relies on pressure vessel/materials engineers to assure integrity
- Allows the authorized inspector to base authorization and approvals on consultations with the pressure vessel/materials engineer

API's practice follows from its underlying concept of adhering to the principles of the ASME Code while allowing flexibility to use engineering judgment

Both codes permit the authorized inspector to be an employee of the owner/operator, but the NB prohibits an employee from approving work performed by the employer unless the authority having jurisdiction (or NB) has given its consent upon review of the owner/operator inspection procedures (Paragraph 301.2d); the API contains no such restriction

Reports, Records, and Nameplates

NB establishes a formal administrative procedure for documenting and recording repairs, alterations, and reratings of pressure vessels

An "R-1" Form, shown in Figure 7.2, must be completed by the organization performing the work (with the exception of routine repair) and submitted to the authorized inspector for approval

Copies of the "R-1" Form are sent to the owner/user, the authority having jurisdiction, and the National Board (alterations only) for permanent record

API requires only that the owner/user maintain permanent records that document the work performed

Both organizations require attaching a new nameplate adjacent to the original nameplate when a vessel is altered or re-rated (refer to Figure 7.3 for an example)

NB also requires attaching a new nameplate to a vessel that has been repaired (with the exception of routine repairs), as shown in Figure 7.4; API has no such requirement

▸ REPAIRS

General Considerations

Repair of a pressure vessel is the work necessary to restore the vessel to a suitable condition for safe operation at the original design pressure and temperature, providing that there is no change in design that affects the rating of the vessel

When deterioration renders a vessel unsatisfactory for continued service, it must be either repaired or replaced

Figure 7.5 provides a "decision tree" that can be used to decide between repair and replacement

Major factors that should be considered in making the decision are shown

It may be necessary to deviate from these steps because of unique local circumstances

CODE AND JURISDICTION REQUIREMENTS

FORM R-1, REPORT OF WELDED REPAIR OR ALTERATION as required by the provision of the National Board Inspection Code

1. Work performed by____

(name of repair or alteration organization) (P.O. no., job no., etc)

(address)

2.

Owner____

(name)

(address)

3. Location of

installation____

(name)

(address)

4. Unit identification:________________________ Name of original

manufacture:____

(boiler, pressure vessels)

5. Identifying nos.: ____

(mfr's. serial no.) (original National Board no.) (Juridiction no.) (other) (year built)

Figure 7.2 R-1 form prescribed by the National Board Inspection Code for repair and alternation of a pressure vessel.

6. Description if work:

(use back, separate sheet, or sketch if necessary)

---

7. Replacement Parts. Attached are Manufacturers' Partial Data Reports properly identified and signed by Authorized Inspectors for the following items of this report:

Figure 7.2 *(continued)*.

(name of part, item number, mfg's. name and Identifying stamp)

8.

Remarks:

This form may be obtained from The National Board of Boiler and Pressure Vessel Inspectors,

1055

Crupper Ave., Columbus, OH 43229

NB-66

Rev. 5

Figure 7.2 (*continued*).

CODE AND JURISDICTION REQUIREMENTS

DESIGN CERTIFICATION

The undersigned certifies that the statements made in this report are correct and that the design changes described in this report conform to the requirements of the National Board Inspection Code.

ASME Certificate of Authorization no.___________ to use the ___________symbol expires

______, 20____

Date ______, 20___ ____signed

____

(name of organization) (authorized representative)

CERTIFICATE OF REVIEW OF DESIGN CHANGE

The undersigned, holding a valid Commission issued by The National Board of Boiler and Pressure Vessel Inspectors and certificate of competency Issued by the state or province of

_____________ and employed by ________________________of

____________________________ has examined the design change as described in this report and verifies that to the best of his knowledge and belief such change complies with the applicable requirements of the National Board Inspection Code. By signing this certificate, neither the undersigned nor his employer makes any warranty, expressed or implied, concerning the work described in this report. Furthermore, neither the undersigned nor my employer shall be liable in any manner for any personal injury, property damage or loss of any kind arising from or connected with this inspection, except such liability as may be provided in a policy of insurance which the undersigned's insurance company may issue upon said object and then only in accordance with the terms of said policy.

Date ______, 20___ Signed _____________ Commissions

____

(Authorized Inspector) (National Board (incl. endorsements), state, prov., and no.)

Figure 7.2 *(continued).*

CONSTRUCTION CERTIFICATE

The undersigned certifies that the statements made in this report are correct and that all

construction and workmanship on this ___________________ conform to the National Board

Inspection Code.

(repair or alteration)

Certificate of Authorization no. _________ to use the ________ symbol expires

______________, 20 ____

Date ________, 20 ____ ____________________________________ Signed

____

(repair or alteration organization) (authorized representative)

CERTIFICATE OF INSPECTION

The undersigned, holding a valid Commission issued by The National Board of Boiler and

Pressure Vessel Inspectors and certificate of competency issued by the state or province of

_______________ and employed by __________________ of ________________ has

inspected the work described in this report on ______ 20 ____ and state that to the best of my

knowledge and belief this work has been done in accordance with the National Board Inspection

Code. By signing this certificate, neither the undersigned nor my employer makes any warranty,

expressed or Implied, concerning the work described In this report. Furthermore, neither the

undersigned nor my employer shall be liable In any manner for any personal injury, property

damage or loss of any kind arising from or connected with this inspection, except such liability

as may be provided in a policy of Insurance which the undersigned's insurance company may

Issue upon said object and then only In accordance with the terms of said policy.

Figure 7.2 *(continued).*

Generally it is more economical to repair a vessel than to replace it, but the primary consideration is integrity and reliability for continued service

Some forms of deterioration, such as creep and hydrogen attack, may indicate that the useful remaining life of the vessel is too short to justify the expense of a repair

*________________________BY________________________

MAWP ________________________psi at________________________°F

(Maximum Allowable Working Pressure)

____

(Manufacturer's Alteration Number, If Used)

____

(Date Altered)

* Insert The World "ALTERED" Or "RERATED" As Applicable

Figure 7.3 New nameplate prescribed by the National Board Inspection Code for a pressure vessel that is altered or repaired.

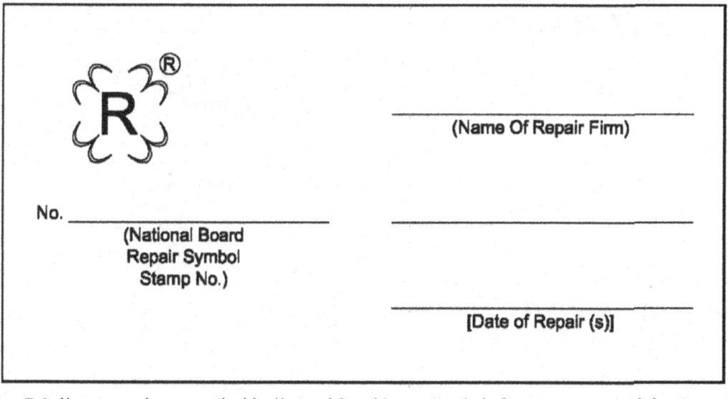

Figure 7.4 New nameplate prescribed by National Board Inspection Code for a pressure vessel that is repaired.

The detection of other forms of deterioration, such as H_2S stress cracking, may indicate that the vessel is not satisfactory for the service environment and the deterioration will recur after repair, thus presenting a continuous maintenance problem

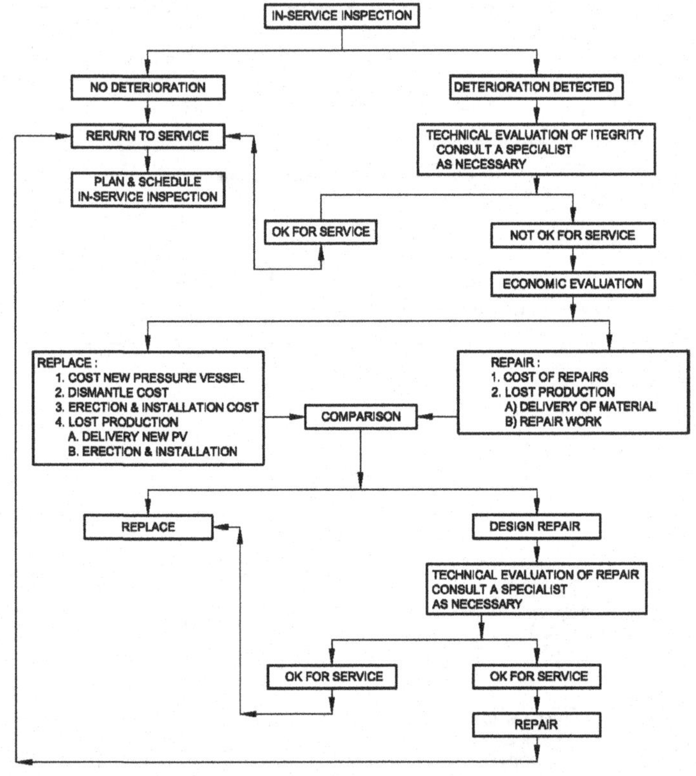

Figure 7.5 Logic tree for repair or replacement of a pressure vessel.

Design of the Repair

Corrosion and cracks at weld joints are the two most common forms of deterioration

Figure 7.6 schematically illustrates three ways in which vessels exhibiting cracks or corrosion can be repaired

Some general approaches to the repair of common forms of deterioration are discussed here, along with the benefits and disadvantages of each

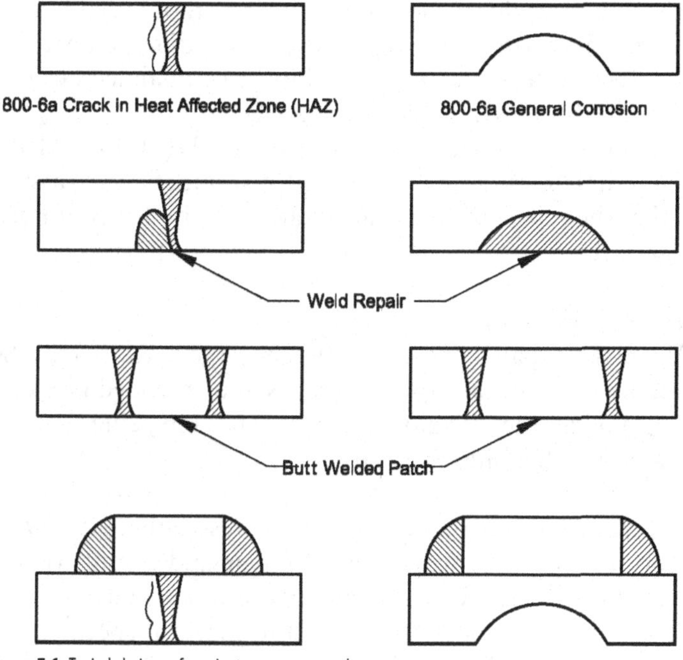

Figure 7.6 Typical designs of repairs to pressure vessels.

Weld Repair

- The simplest repair of cracks consists of removing a crack by gouging or grinding and filling the groove with weld metal to restore the shell to the minimum required thickness, plus corrosion allowance
- A corroded area can be ground smooth and free of corrosion scale and then restored to the minimum required thickness plus corrosion allowance by weld build-up
- The ground area should be examined by Magnetic Particle testing (MT) or Dye Penentant testing (PT) to be certain that all of the cracked or Magnetic Particle testing otherwise damaged Dye Penentant testing material has been removed

Welding procedures should be similar to those used for the original construction of the vessel, including preheat and post-weld heat treatment, and must be qualified according to the ASME Code, Section IX

The entire weld repair should be examined by UT or RT for internal flaws and by MT or PT for surface flaws to assure that the repair will have adequate integrity for continued service

Insert Patch (Flush Patch)

When the repair covers a relatively large area, it may be more economical to repair cracks and corroded areas by removing the deteriorated area and replacing it with a butt-welded insert patch

If the deterioration area is very large, it may be beneficial to replace the entire component of the vessel; there is usually no difference between a weld repair and a butt-welded patch with regard to the integrity and reliability

Patch should be made from the same material specification and grade as the vessel components

If the material is obsolete, the patch should be made from a material that has similar strength, C_V impact toughness, and welding characteristics

Procurement of the proper material with the correct thickness can add to the "lead time" for making a repair and should be planned as far in advance as possible

Insert patches should be formed to the radius of the shell component that they will be used to repair

Square or rectangular patches should have rounded corners with a radius that is at least four times the thickness

Care should be taken to obtain as good as possible fit-up of the butt patch with the actual size and shape of the opening in the shell into which it will be inserted, and this fill-up should be maintained during welding through the use of temporary attachments (clips) to the patch and shell

Hot Cracking of the Weld Metal on Cooling

Can be a problem with butt patches because of the relatively high restraint imposed by the surrounding vessel shell

Preheating can help alleviate this problem

Distortion (flattening) of the patch can also occur

Shrinkage of the weld metal upon cooling tends to pull the patch flat with respect to the matching radius of the vessel shell, and this can be of concern because it will change the stressor developed in the patch

The use of "strong backs" temporarily attached to the patch can minimize the distortion

Every effort should be made to maintain the patch and adjoining vessel shell within the ASME Code tolerances for out-of-roundness, to make certain that the stresses developed in the patch do not exceed those permitted by the code

All clips and strong backs should be removed after butt welding of the patch to the shell is completed by cutting above the attachment welds

After their removal, the locations where they were attached to the patch and adjoining shell should be ground flush and smooth

NDE Examination

The butt weld should receive full RT or UT examinations (regardless of the examination performed during construction) to assure the integrity of the repair

Full-coverage examination is important because of the possibility of hot cracking of the weld metal

MT and PT examinations should be made on all locations where the temporary attachments (clips and strong backs, etc.) were removed from the patch and adjoining vessel shell

Fillet-Welded Patch

Its use is sometimes thought of as an expedient way to avoid encountering lengthy out-of-service periods

Does not conform to the definition of a repair because it constitutes a change in the design of a pressure-containing component

Thus, it should be handled as an alteration

API 510, Paragraph 8.1.5.1.2, recognizes the utility of this type of repair and provides guidelines for its design

API 510 Requirements

A fillet-welded patch should provide equivalent integrity and reliability as a reinforced opening

The primary membrane stress in the lap patch should not exceed the maximum allowance design stress for the material given in the ASME Code, and the elastic strain in the patch should not result in the fillet welds being stressed above their maximum allowable stress

Joint Efficiencies

Major constraint in designing a lap patch is to meet the joint efficiency permitted by ASME Code for lap welds (Table UW-12)

It is not possible to examine adequately, by RT or UT, the fillet welds for a lap patch to assure the absence of flaws; thus, the efficiency of a lap-welded joint is limited to 0.50

Because of the low joint efficiency, it is almost impossible to design a lap patch that will provide integrity equivalent to that of the original design and conduction

Fillet-welded patches are therefore not recommended for making permanent repairs

Planning and Approval

Both the NB (Paragraph R-301.1) and API 510 (Paragraph 5.1.1) require obtaining authorization for making a repair from the authorized inspector before the work is initiated, except for routine repairs when the authorized inspector has given prior approval

Examples of routine repairs are as follows:

- Weld build-up of corroded areas
- Application of corrosion-resistant weld overlay
- Addition of non-pressure-containing attachments when PWHT is not required
- Replacement of flanges

Authorization for making a repair that is not routine is obtained from an authorized inspector by preparing and submitting a repair plan

The repair plan should be prepared by an engineer at the facility/plant in consultation with a maintenance coordinator

The repair plan should include the following information:

- Areas of vessel to be repaired
- Repair procedures to be used for each area specifying the following:
  - Preparation for repair (removal of deterioration)
  - Materials
  - Welding procedures
  - NDE of repairs

Repairs that will be made by a contractor should be discussed with the contractor to obtain agreement with the plan before it is submitted to the authorized inspector

Under "emergency repairs," the repair can be initiated prior to submitting the plan to the authorized inspector, but complete documentation should be preserved and submitted for the inspector's acceptance as soon as possible; the vessel cannot be returned to service unless the authorized inspector has accepted the repair

Organization Making the Repair

The NB (Paragraph R-404) requires the organization performing a repair to have either a certificate of authorization (COA) from the NB for the use of an "R" stamp or a COA from ASME for the use of a "U" stamp

API 510 (Paragraph 1.2.13) also accepts an organization having an ASME "U" stamp as qualified to make repairs, but it makes no mention of an NB "R" stamp

API 510 permits owners/operators to repair their own vessels in accordance with their requirements and to have repairs made by contractors whose qualifications are acceptable to them

All repairs that are not routine should be performed by an organization that has a valid "U" stamp, regardless of code or jurisdiction requirements that might permit repair by other organizations

Repair of Materials

Both NB (Paragraph R-305) and API 510 (Paragraph 5.2.7) require that the materials used for repair must conform to one of the specifications in ASME Code, Section II

Materials should be the same as those used for the original construction; if this is not possible, alternate material selection should be discussed with materials engineers

Replacement Parts

A repair can involve replacing a deteriorated part with a new part of the same design that is manufactured in a shop

Manufacturing a replacement part generally requires welding

If the ASME Code requires inspection of the weld joints by an authorized inspector, the NB requires the replacement part to be manufactured by an organization that has an ASME certificate for a "U" stamp (Paragraph R-307.1c)

A "U" stamp with the word "part" is applied to the part when the authorized inspector accepts it

Replacement parts do not require inspection by an authorized inspector and are not required to be manufactured by a holder of an ASME COA (Paragraph R-307.1b)

API 510 requires replacement parts to be manufactured according to the principles of the ASME Code, but it has no requirement concerning the qualifications of the manufacturer (Paragraph 8.1.3)

It is recommended that all replacement parts be manufactured by an organization that has a COA from ASME for the use of a "U" stamp

Repair Welding

NB Requirements (Paragraphs R-302.1 and R-302.2)

Qualification of all welding procedures according to ASME Code

Welders to pass a welder performance qualification for each welding procedure used

Repair organization to make the records of procedure and performance qualifications available to the authorized inspector before actual repair welding begins

API 510 Requirements (Paragraph 8.1.6.2)

Repair organization is to qualify all welding procedures and welders used for a repair according to the principles of ASME Code, Section IX

"According to principles" allows more flexibility for deviating from a welding procedure acceptable to the ASME Code when necessary to expedite a repair

Welding procedures should not deviate from the code unless a materials engineer has reviewed them

Post-Weld Heat Treatment (PWHT)

Can be a difficult aspect of the repair and, when improperly performed, can cause additional damage to the vessel

Repair welds should receive the same PWHT used for the original construction whenever possible

PWHT of a repair weld is especially important when it was specified for the original construction of the vessel to prevent stress corrosion cracking

PWHT of a repair weld is most often accomplished by the local application of heat to the repaired area while the remainder of the vessel is at ambient temperature

During PWHT, high thermal stresses that can damage the vessel may develop because of severe temperature gradients and restricted thermal expansion

Nozzles, head-to-shell weld joints, attachment welds for vessel supports, piping connections, and internal components are particularly vulnerable to damage

The vessel must be free to expand when the local area is heated, and efforts should be made to keep temperature gradients less than 100°F per foot along the surface and 100°F per inch through the thickness at temperatures above 400°F

When it is not possible to perform a local PWHT within these guidelines, the risk of damage to the vessel should be carefully evaluated and alternatives for repair without PWHT should be considered

Both the NB (Paragraph R-303.2.2) and the API (Paragraph 8.1.6.4.2) permit substituting a temper bead (or half bead) welding procedure for PWHT for the repair of carbon steel vessels

Neither organization requires a separate qualification of this welding procedure to demonstrate that the weld metal and heat-affected zones of the repaired vessel will have the properties required to assure adequate integrity for continued service (minimum strength, maximum hardness, and C_V impact toughness)

ASME Code, Section IX, contains superior requirements for qualifying and performing this type of repair weld

These procedures should not be used for the repair of a vessel unless they are discussed with a material engineer

Inspection and Hydrotest

Inspection of Repairs

Both the NB (Paragraph R-301.2) and API 510 (Paragraph 5.8.1.1) require the acceptance of repairs to a pressure vessel by the authorized inspector before the vessel is returned to service

The authorized inspector will normally require performing all of the NDE examinations for the repair that were required by the ASME Code during original construction

Alternative NDE methods can be proposed (such as the substitution of UT for RT) when it is not possible, or practical, to use the NDE method that was used during construction

All repair welds to vessels should be subjected to essentially full coverage NDE, in view of the more difficult working conditions that will usually be encountered for repairs compared to the favorable conditions in a fabrication shop

UT is an entirely acceptable NDE method for verifying the quality of welds and does not involve the hazards and obstruction of other work associated with RT

Hydrotest after Repairs

Neither the NB (Paragraph R-308.1) nor API 510 (Paragraph 5.8.1.1) makes it mandatory to perform a hydrostatic pressure test following the repair of a pressure vessel, but agreement from the authorized inspector is required for it be waived

The purpose of the hydrotest in the ASME Code is to detect gross errors in the design or major flaws in the construction of a new vessel

Repair of a vessel restores it to a satisfactory condition without any change in design; therefore, there is no need to verify the design of the repaired vessel

Full-coverage NDE of all repairs will detect much smaller flaws than those that could cause failure during a hydrotest and will therefore provide a greater assurance of the quality of the repair than a hydrotest

In-Service Inspection after Repairs

Should be planned and scheduled after it has been returned to service to assure that the repair is providing sufficient integrity

Especially important when a repair has been made by deviating from the original construction requirements or from the Code of Construction

NDE methods should be used to detect deterioration of the vessel that necessitated the repair

Approval of Repairs, Documentation, and Nameplate

NB Requirements (Paragraph R-402)

Repair organization to document the repair by completing an R-1 Form that is submitted to the authorized inspector for approval (see Figure 7.2)

Subsequent to obtaining approval of the R-1 Form, the repair organization must attach a new nameplate

Nameplate is stamped with an "R" if the repair organization has a COA from the NB

Repair organization cannot stamp the nameplate with a "U," despite using its COA from the ASME to qualify it for making the repair

ASME only permits the "U" stamp for the design and construction of a new vessel

Completion of an R-1 Form and attachment of a new nameplate may not be required for routine repairs, dependent on the following:

Authorities having jurisdiction

Approval of the authorized inspector

API 510 Requirements (Paragraph 8.1.2.1)

Requires documentation of repairs to be kept as permanent records, but does not prescribe using a standard form

Does not require attaching a new nameplate to a repaired vessel

A new nameplate should not be attached to a vessel after a repair unless the authorities having jurisdiction mandate following the National Board (NB)

The original nameplate of the vessel has the primary purpose of permanently displaying the following:
- Maximum Allowable Working Pressure (MAWP)
- Temperature rating

A repair does not change the rating of the vessel; thus, a new nameplate is unnecessary, unless required by the authorities having jurisdiction

ALTERATION

General Considerations

Alteration of a pressure vessel

Physical change to any component that affects pressure-containing capability

Can change the MAWP and temperature rating of a vessel from that given on the original nameplate with a "U" stamp

Alterations can be designed so as not to affect the original rating of a vessel, provided the operating pressure and temperature are not changed

Alterations are usually made to accommodate the following:
- Changes in process design
- Installation of new nozzles
- Changes in internal components

The effect that the design loads on the new internals (pressure drop, static weight, liquid head, etc.) have on the stressor in the vessel shell should be calculated to determine if the MAWP of the vessel has to be changed

Planning and Approval

NB (Paragraph R-501) requires all alterations to conform to the ASME Code

API 510 (Paragraph 8.1.1) requires adhering to the principles of the ASME Code

Wording allows more flexibility for designing alterations when it is not advisable or practical to conform to the code

Both the NB (Paragraph R-301.1.2) and API 510 (Paragraph 8.1.2.1) require authorization from the authorized inspector prior to initiating an alteration

Authorized inspector will do the following:

- Verify that the design of the alterations and calculations have followed the ASME Code criteria
- Determine that acceptable materials will be used
- Assure that the weld procedures and welders are properly qualified

API 510 requires the inspector to consult with an experienced materials engineer before giving authorization to proceed with the alteration

Organization Making Alterations

NB (Paragraph R-505) requires that an organization performing an alteration has an ASME COA covering the scope of work involved

API 510 does not contain specific requirements for an organization performing an alteration

Presumably, the same requirements would apply as for a repair organization

Alterations can be designed by qualified pressure vessel and materials engineers, but it is recommended that only organizations holding an ASME COA perform the work on the vessel

Materials, Replacement Parts, Welding, PWHT, and Inspection

The requirements for the alteration of pressure vessels concerning materials, replacement parts, welding, PWHT, and inspection are identical to those for repair

Hydrotest after Alterations

Hydrotesting alterations is a mandatory requirement of the NB (Paragraph R-308.2)

API 510 (Paragraph 5.8.1.1) states the following:

Hydrotesting is normally required after an alteration

Permits waiving the hydrotest after consultation with a materials engineer if superior designs, materials, fabrication procedures, and inspections are used

A hydrotest should be performed after an alteration whenever possible

An alteration, by definition, changes the design of at least one component of the vessel shell, and the validity of the design changes cannot be verified by comprehensive inspection

An alteration differs significantly from a repair, which does not involve design changes

The pressure for the hydrotest should be the minimum test pressure required by the ASME code for the nameplate design pressure and temperature

The test pressure, after alteration, will normally be lower than the recommended hydrotest pressure for new vessels, as it is likely that some of the original corrosion allowance will have been consumed during service before the alteration is made

Hydrotesting pressure vessels that have been altered by the installation of a new nozzle requiring reinforcement is sometimes accomplished by welding a cap to the inside of the vessel shell covering the nozzle

This circumvents preparing the entire vessel for hydrotest by providing for a "local hydrotest"

However, a local hydrotest will not develop the same stresses in the nozzle reinforcement and the vessel shell component surrounding the opening, as would be developed by hydrotesting the entire vessel

The cap will effectively change the shape of the vessel shell component to which it is welded and have a significant effect on the stresses developed in that component by internal pressure

Thus, a local hydrotest is not a valid verification of the design of an alteration, and this practice is not recommended

Approval of Alterations, Documentation, and Nameplate

NB Requirements (Paragraph R-502)

Requires that the organization performing the alteration prepare an R-1 Form, which must be submitted to the authorized inspector for approval

The organization performing the repair must then attach a new nameplate that displays the design pressure (or MAWP) and temperature for the altered vessel

Approval of the alteration and attachment of the new nameplate must be obtained from the authorized inspector before the vessel is returned to service

API 510 Requirements (Paragraph 7.8.2 (c))

Requires that the documentation of alterations to pressure vessels must be kept as permanent records, but it does not prescribe using a standard form (Paragraph Appendix D)

Approval of an alteration by the authorized inspector is required before the vessel is returned to service, but attachment of a new nameplate is not mandatory unless the design pressure (or MAWP) and temperature are changed by the alteration

▸ RE-RATING

General Considerations

Re-rating a pressure vessel consists of changing the design pressure (or MAWP) or temperature from those displayed on the vessel's nameplate

Re-rating usually *does not* involve a physical alteration of the pressure-containing capability of the vessel, but it can be required by alterations that are not designed for the original design pressure or temperature

Re-rating is most commonly necessitated by the following:

- A change in operating conditions for the process
- Deterioration (i.e., the occurrence of corrosion or cracking) that affects vessel integrity and reliability for the original design pressure and temperature, so that a repair cannot be economically justified

Organization Making Re-Rating

NB Requirements (Paragraph R-503)

Requires the re-rating of a pressure vessel to be performed by the original manufacturer whenever possible

The re-rating can be performed by a registered professional engineer if the re-rating cannot be obtained from the manufacturer

API 510 Requirements (Paragraph 8.2.1)

Permits either the original manufacturer or an experienced engineer employed by the owner/operator to perform the re-rating

Only engineers with appropriate experience with pressure vessel design, fabrication, and inspection should perform re-ratings

A consultant retained by the owner/operator is also acceptable

Calculations

Re-rating a pressure vessel requires making calculations for every major pressure-containing component (i.e., shell, heads, nozzles, reinforcements, flanges, etc.) to verify that they will be adequate for the new design pressure and temperature

The effect of all internal and external loads on the vessel shell must be considered in the calculations for re-rating

Therefore, re-rating involves repeating all the calculations that were made for the original design of the vessel for the new design pressure and temperature

It can be thought of as designing a pressure vessel in reverse

Instead of calculating the minimum required thickness of each shell component for the prescribed design pressure and temperature, the calculations are made to determine if the actual thickness of each shell component is adequate for the re-rated pressure and temperature

Both NB (Paragraph R-503) and API 510 (Paragraph 8.2.1) require making calculations according to the ASME Code

Decrease in Pressure

Re-rating of a pressure vessel for a lower pressure is usually required if

The operating temperature is increased for new process conditions

Corrosion has reduced the remaining wall thickness below the minimum required thickness for the original design conditions

Increase in Pressure

Re-rating of a pressure vessel for a higher pressure can usually be accomplished only under the following conditions:

The operating temperature is decreased

Thickness measurements of all pressure-containing shell components indicate that the original corrosion allowance was greater than necessary for the actual corrosion experienced

Increase or Decrease in Temperature

An increase in the temperature will almost always require decreasing the pressure, unless the new temperature remains below 450°F

A decrease in temperature will almost always permit an increase in pressure, unless the original temperature was 450°F or below

Re-rating for a lower temperature should never be allowed to violate the rules in the ASME Code for low-temperature operation

Essential to check the vessel being re-rated for compliance with the current rules of the code for low-temperature operation when the new temperature will be 120°F or below

This may be difficult to do when the vessel is old and the materials used for construction are now obsolete

Under these circumstances, it may be necessary to cut samples from the vessel for C_V impact testing to perform a satisfactory re-rating

Information Required

A thorough inspection should be made to assure that the vessel is in satisfactory condition for the new pressure or temperature

It is especially important to determine the minimum remaining thickness of every pressure-containing component of the vessel shell and to detect any cracks that may have developed during service

This will usually require more NDE than normally performed during a routine in-service inspection

Approval of Re-Rating, Documentation, and Nameplate

NB treats the re-rating as an alteration with respect to the requirements for preparation of an R-1 Form, approval by an authorized inspector, and attachment of a new nameplate displaying the new pressure or temperature

API 510 also requires approval of the re-rating by an authorized inspector and attachment of a new nameplate

The new nameplate should be considered mandatory because the pressure or temperature for the re-rated vessel differs from that displayed on the original nameplate

Case Study: Major Vessel Shell Repair

Background

Because of severe fouling internally in two shell and tube heat exchangers, it became necessary to split the shells along their longitudinal axis to remove the bundles

Many attempts had been made to avoid this extensive work, but it became the only solution

After removing the tube bundles from their shells, the task of repairing the shells became major steps in reconditioning the heat exchangers

This case study covers the repair steps necessary to restore the shells

The shells were equipped with body flanges, which were cut around their entire circumference, and the shells were cut their entire length

The tube bundles were extracted and were refurbished while the exchanger shells were being repaired

Stabilizing Tabs

Once the tube bundle has been removed, the shells must be cleaned before repairs

After cleaning, a shell is held in place with chains, and then the internal circumference is set back to its original dimension

Figure 7.7 shows stabilizing tabs that were welded on the shell to hold the shell's circumference in place for the welding

More stabilization will be needed and was in the form of strong backs as shown in figures 7.10 and 7.11

Figure 7.7 Stabilizing tabs hold the shell in place for major shell repair.

Figure 7.8 Edges of the seam have been flame cut and will need to be ground to a bevel.

Strong backs are used to minimize the distortion that will be a concern during the welding processes

If strong backs are not used, the area of the shell along the long seam will become a flat spot for its length, which code rules do not permit

Backing Bar

Because of the large weld joint opening, the shell requires a flat backing bar to support the weld metal

Welding is from the inside and the backing bar on the outside, which makes removal of the bar after welding the initial passes much easier

As shown in Figure 7.8, the edges of the seam have been flame cut and will need to be ground to a bevel before welding can begin

Strong Backs

Strong backs are welded to do the following:

- Hold the shell in place as welding progresses

Figure 7.9 Welding from outside is faster.

- Control shrinkage that might cause distortion as the weld metal solidifies

Strong backs may be placed on the inside or outside of the shell, depending on which side the initial weld passes will be made

In Figure 7.9 manual welding of this shell's long seam has commenced and will continue over 12-hour shifts

Internal Strong Backs

As shown in Figure 7.10, quarter moon strong backs maintain the inside circumference

In this repair, welding was performed from the outside first

After completion of the outside weld, the strong backs will be removed and the internal weld joint will be ground to smooth sound metal and the weld completed from the inside

A total of three vessels that were cut open to remove the bundles were repaired in this manner

External Strong Backs

In cases where it is desirable or necessary to make the weld from the inside first

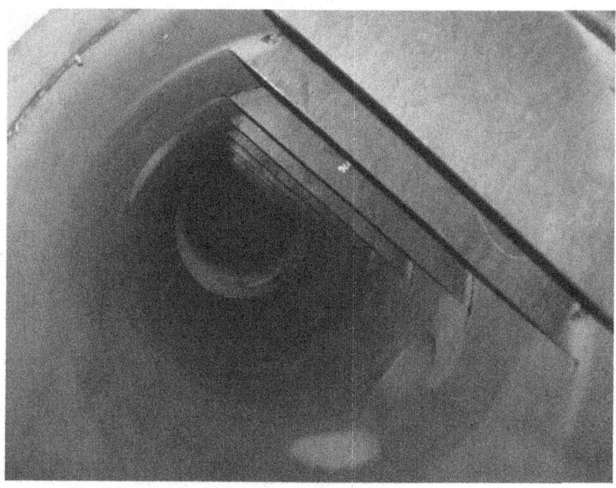

Figure 7.10 Quarter moon strong backs maintain the shell circumference.

Strong backs are welded on the outside of the shell (Figure 7.11)

These strong backs were the remains of the steel plate used to make the internal strong backs shown in Figure 7.10

Because two identical vessels were being repaired, one was welded from the inside and the other from the outside, eliminating wasting the strong back plate.

Normally the shell is rolled into a nearly flat position to make the welding easier

A backing bar was used as the weld prep groove has a wide spacing; the strong backs and bar will be removed to complete the weld from the outside

Flat Metal Backing Strip (Bar)

Routinely, first weld passes are performed using a manual or semiautomatic process such as shielded metal arc or gas metal arc welding (SMAW/GMAW)

Figure 7.11 External strong backs.

If this metal backing strip were not used, it would be difficult to bridge the molten weld metal across such a wide opening

The balance of the weld can then be made using the GMAW process for increased speed or SMAW as appropriate

The strong backs will be removed once the shell has had enough weld metal deposited to hold its dimension stable (Figure 7.12)

Rabbit (Template)

After welding has been completed and the strong backs removed, a template is passed through the shell to check for diameter, roundness, and obstructions.

Any weld reinforcement (weld cap) is ground flush in exchanger shells to provide the maximum clearance for the tube bundle

The template used is known as a "rabbit" (Figure 7.13)

This is done to make sure the shell's inside diameter provides for the outside diameter of the exchanger's tube bundle plus adequate but not excessive clearance

Figure 7.12 Flat metal strip use as weld backing.

Figure 7.13 A template is pushed through the shell to check for obstructions and its dimensions.

Reattaching the Body Flange

After the shell has been welded, the shell body flange is fitted to the shell cylinder for welding

The shell cylinder is stabilized by the completion of its longitudinal weld seam and is made round to meet ASME Pressure Vessel Code specifications

The difference between the maximum and minimum inside diameters at any cross section shall not exceed 1% of the nominal diameter at the cross section under consideration

Shell flange is attached using the same approach as new construction; in Figure 7.14 it has to be attached using small welds in preparation for welding by the submerged arc process

Submerged Arc Welding Process (SAW)

The body flange in Figure 7.14 will be attached using the SAW process

Figure 7.14 Shell body flange is fitted to the shell for welding.

Figure 7.15 shows the process as used to weld the longitudinal weld seam of a vessel cylinder

For welding the body flange shown in Figure 7.14, the exchanger shell will be mounted on an automatic rolling machine and the SAW process used to weld the entire circumference

By employing several revolutions of the assembly, completing this process is relatively fast, accurate, and can have excellent control of welding heat input thereby minimizing distortion

Post-Weld Heat Treatment (PWHT)

Post-weld heat treatment is one of many types of heat treatment such as the following:

- Annealing to soften metal for forming or machining
- Normalizing to provide uniformity in grain size
- Quenching to harden ferrous metals

Figure 7.15 Example of the submerged arc welding process.

It can be required in the following circumstances:

- It is required by the Code of Construction to relieve the locked-up stresses produced by the welding process in materials based on the thickness/material combinations
- In welding operations it can be required to relieve the weld metal and heat-affected zone (HAZ) stresses created during welding for service environment reasons without regard to thickness

Locked-up stress can result in stress-corrosion cracking, distortion, fatigue cracking, premature failures, and accelerated corrosion at the weld and its heat-affected zone

Additionally when the stress from an applied load, such as internal pressure in an exchange shell, is added to the residual stress left by the welding process, the total tensile load can be far greater than that induced by the pressure loading alone

The exchanger shell repaired in the previous shell repair section was originally post-weld heat treated and was required to receive PWHT after the repairs

Heat treatment can be performed by placing an entire exchanger shell in a heat-treating oven or by performing it locally on selected welds

Often it is more efficient to simply move the entire exchanger's vessel into an oven, as shown in Figure 7.16

This heat-treatment oven can accommodate fairly large and thick exchanger shells

Ovens such as this one are normally used when local heat-treatment methods may not be sufficient or practical

This is only one of the ovens in this repair shop; this shop has two ovens as well as local heat-treatment capabilities

In Figure 7.17 the oven adjacent to the oven of Figure 7.16 is loaded with a exchanger shell that is to be heat treated

Figure 7.16 Medium-capacity heat-treatment oven.

The shell has only been loaded at this stage; it still needs to be equipped with thermocouples for monitoring its temperature during the heat-treatment operation

It will require more than one thermocouple; normally a minimum of two are required, even for small shells

PWHT minimum temperature and time at temperature (referred to as soak time) are usually obtained from the Code of Construction; for other forms of heat treatment, the criteria are based on good engineering practices

Heat treatment temperatures should be precisely controlled; Figure 7.18 shows the control room for the ovens pictured in Figures 7.16 and 7.17

These heat-treatment ovens are natural gas fired; two sets of controls shown in Figure 7.18 control the flow of gas to the oven's burners, modulating the rate of firing in order to control the temperature inside the oven to a desired preset temperature

Figure 7.17 Exchanger shell being prepared for PWHT.

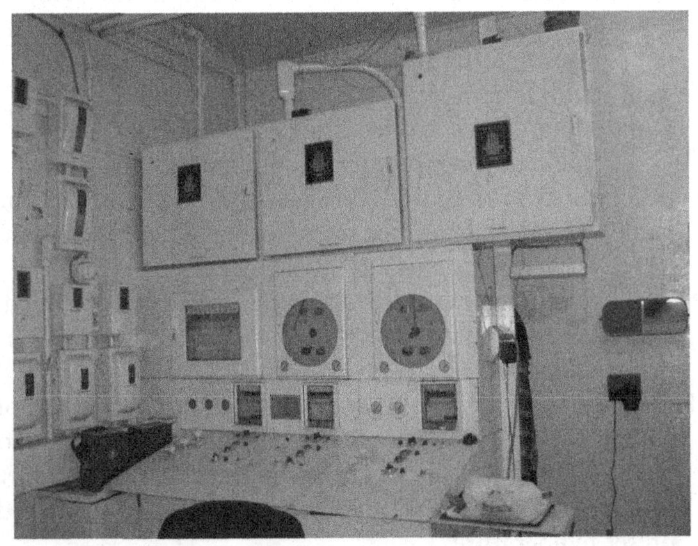

Figure 7.18 Heat-treatment oven control room.

The exchanger metal temperature is sensed by the thermocouples directly attached to the exchanger shell

Notice also in the figure that above the controls shown, at about eye level standing, are two chart recorders

The chart recorder is operating the entire time during heat treatment, and the chart becomes a permanent record and is shipped with the heat-treated equipment to the customer

The heat-treatment process is determined by the applicable code when done for stress relieving

Codes may have listed parameters such as the following:

- Maximum temperature of the oven when the part is placed inside the oven
- Rate of increase in temperature
- Required minimum holding temperature
- Minimum time at temperature
- Cool-down procedure
- Just to name a few requirements, consider these examples; ASME Section VIII, Division 1, Paragraph UCS 56, requires the following:
- The part cannot be placed in the oven at a temperature greater than 800°F (425°C)
- Above 800°F (425°C), the part must not be subjected to a rate of heating greater than 400°F (222°C) per hour based on thickness
- Other requirements, such as differential of temperature over a given length of the part, flame impingement, and finally the rate of cooling acceptable to the code, are issues that must be considered when a manufacturer or a repair concern performs heat-treatment operations

Index

D

M

N

O

P

T

Z